Biofuels: implications for the feed industry

Biofuels:
implications for
the feed industry

edited by:
Jannes Doppenberg
Piet van der Aar

Wageningen Academic
P u b l i s h e r s

ISBN 978-90-8686-043-2

First published, 2007

Wageningen Academic Publishers
The Netherlands, 2007

Table of contents

Biofuels: implications for the feed industry

P.J. van der Aar
Schothorst Feed Research B.V.

> *The increasing global demand for energy and the consequences of the use of fossil energy on global heating and production of greenhouse gasses, have generated an accelerating interest in energy production from sustainable sources.*
>
> *One of the possible alternatives is the production of energy from biomass.*

During the last 5 years many initiatives to produce energy from biomass have emerged. This development has various consequences for the feed industry, like:
- an increasing competition between food, feed, fuel and other industrial applications for the use of agricultural products;
- a change in the supply and the type of feedstuffs worldwide;
- more by-products become available for feed production and will compete and/or replace at least partially traditional feed ingredients.

The competition between feed, food and fuel for agricultural products will put a tremendous pressure on the agricultural community to meet the demands. Already during the last 5 years the use of cereals globally has been larger than the production, resulting in diminishing stocks. The expected extension of the production of biofuels will enlarge this challenge. Since areas suitable for agricultural production are limited, the production per acreage has to increase. Policymakers will have to support the agricultural community in the (re)development of production oriented knowledge en expertise. It will need a change in the current policies on agricultural research.

The initiatives taken to produce biofuels are mainly based on the production of biodiesel and ethanol. These are the first generation products. In both types of production the energy component of the crops are used. This results in a lower availability of both starch and vegetable fats for feed. The residues will be protein and fibre rich. These will become available in larger quantities for the feed industry. It can be expected that the price of feeds will be determined by the energy content of these feedstuffs, whereas protein will become relatively less valuable. The protein rich by-products, like rapeseed oil meal and DDGS, will directly compete with traditional protein sources like soybean meal.

The diversity of production processes from which these by-products are derived causes a wide variation in the quality of the products and thus in the value as animal feed. Not only the chemical composition is affected by the production process, also nutrient availability will vary. The feed industry will need reliable methods to evaluate these raw materials. It is the core competence of Schothorst Feed Research to develop the methods into practical applications and table values for the feed industry.

Not only will the nutritional quality vary. The variation in processes also means that the critical points regarding food safety may vary. Therefore producers of these by-products should realise that food safety should be an integrated part of the production of biofuels.

With this conference we would like to show our commitment to the feed industry that we have a continuous focus on developments relevant to the feed industry, especially on raw material quality, nutritional values and feed and food safety. We will monitor the next generations of techniques of energy production from biomass closely, and study the consequences for the feed industry. We expect that it will be a major theme in our activities for the next decade.

Development of international sustainable biomass and biofuel markets and trade

M. Junginger, A. Faaij, F. Rosillo-Calle and J. Woods
Copernicus Institute, Utrecht University, The Netherlands

Abstract

Rapidly developing international bio energy trade may develop over time into a 'commodity market' which can secure supply and demand in a sustainable way; sustainability being a key factor for long-term security. It is clear that on a global scale and over the longer term, large potential biomass production capacity can be found in developing countries and regions such as Latin America, Sub-Saharan Africa and Eastern Europe. If indeed the global bioenergy market is to develop to a size of 400 EJ over this century (which is well possible given the findings of recent global potential assessments). This creates important future opportunities for such regions, given the expected increased role of bioenergy within the world's energy supply. Consequently, this poses the fundamental question of how these potential major producers and exporters of bioenergy can benefit from the growing global demand for bioenergy in a sustainable way, i.e. that bioenergy exports can contribute to rural development, benefit local communities and can be an integral part of overall development schemes, including the existing agricultural and forestry sectors. In this paper the links between international bioenergy trade, development and how sustainable bioenergy production could be realized are explored. Drivers, barriers and future potentials for international bioenergy markets are discussed and socio-economic implications for possible exporting countries are identified. By doing so, several key opportunities and issues for the developing international bioenergy markets and their possible impacts on developing and rural regions are raised that should be taken into account by policy makers, market parties, international stakeholders and other key stakeholders. Summarizing; although international bioenergy trade and markets are developing very rapidly and the future looks bright given market demand and potential supplies, also many barriers occur that can disturb or at least slow down a sound development of such markets. Also, there are important concerns about competition for land that may conflict with food production, water resources and biodiversity protection. Although biomass production may well provide a crucial strategy to enhance sustainable land-use management, negative developments should be avoided, e.g. by clear standards and best practice guidelines for (the design of) biomass production systems and their integration in agricultural areas.

Keywords: biomass trade, sustainability criteria, logistics, biofuels, biomass potentials, environmental and socio-economic impacts

The growing role of biofuels: opportunities, challenges and pitfalls

We give an overview of biomass production potentials, and the importance of technological development and perennial crops to utilise this potential. Second, we take a look at some of the most critical issues for developing large-scale biomass for energy production, and indicate at the same time how these issues may be avoided or solved. Third, we provide an overview of on-going developments to ensure the 'greenness' of biomass by developing safeguards, e.g. via sustainability criteria and certification schemes for bioenergy. Finally, we highlight the importance of sustainable international bioenergy trade as a major driver to develop biomass potentials.

What are the perspectives for producing biomass for energy

In principle we categorise biomass into three categories: energy crops on current agricultural land; biomass production on marginal lands; and residues from agriculture and forestry, dung and organic wastes. As we show below, we estimate that globally, these categories may supply 200 EJ, 100 EJ and 100 EJ respectively.

   Clearly, biomass production requires land. The potential for energy crops therefore largely depends on land availability, which must account for growing worldwide demand for food, nature protection, sustainable management of soils and water reserves and a variety of other environmental services. Given that a major part of the future biomass resource for energy and materials depends on these intertwined, uncertain and partially policy dependent factors, it is impossible to present the future biomass potential in one simple figure. A review of the literature on future biomass availability carried out in 2002 (17 studies in total) revealed that no complete integrated scenario assessments were available (Berndes *et al.*, 2003). These studies include those by IPCC, US EPA, World Energy Council, Shell, and Stockholm Environmental Institute, and arrived at varying conclusions on the possible contribution of biomass to the future global energy supply (e.g. from less than 100 EJ yr^{-1} to above 400 EJ yr^{-1} in 2050). Table 1 provides a summary of

Table 1. Overview of the global potential of bioenergy supply on the long term for a number of categories and the main pre-conditions and assumptions determining these potentials[a].

Biomass category	Main assumptions and remarks	Potential bioenergy supply up to 2050 (EJ/yr)[b]
Energy farming on current agricultural land	Potential land surplus: 0–4 Gha (Most studies find 1-2 Gha). A large surplus requires intensive agricultural production systems (i.e. modernisation of all aspects). When this is not feasible, the bioenergy potential could be reduced to zero. On average higher yields are likely because of better soil quality: 8-12 dry tonne/ha/yr are assumed.	0–700 (100–300)
Biomass production on marginal lands	On a global scale a maximum of 1.7 Gha could be involved. Low productivity of 2–5 dry tonne/ha/yr. The supply could be low or zero due to poor economics or competition with food production.	0–150 (60–150)
Residues from agriculture	Potential depends on yield/product ratios and the total agricultural land area as well as type of production system. Extensive production systems require re-use of residues for maintaining soil fertility. Intensive systems allow for higher utilisation rates of residues.	15–70
Forest residues	The sustainable energy potential of the world's forests is unclear. Part is natural forest (reserves). Low value: figure for sustainable forest management. High value: technical potential. Figures include processing residues.	0–150 (30–150)
Dung	Use of dried dung. Low estimate based on global current use. High estimate: technical potential. Utilisation (collection) on longer term is uncertain.	(0)5–55 EJ
Organic wastes	Estimate on basis of literature values. Strongly dependent on economic development, consumption and the use of bio-materials. Figures include the organic fraction of MSW and waste wood. Higher values possible by more intensive use of bio-materials.	5 – 50 +

Table 1. Continued.

Biomass category	Main assumptions and remarks	Potential bioenergy supply up to 2050 (EJ/yr)[b]
Total	Most pessimistic scenario: no land available for energy farming; only utilisation of residues. Most optimistic scenario: intensive agriculture concentrated on the better quality soils. (In brackets: more average potential in a world aiming for large scale utilisation of bioenergy).	40 – 1,100 (250 – 500 EJ)

[a]The overview is based on review of 17 studies and Faaij *et al.* (2000), Smeets *et al.* (2004) and Hoogwijk *et al.* (2005).
[b]Note that traditional use of dung as fuel should be discouraged. The dung potentials shown here mainly stem from intensive agriculture, which offers opportunities for fermentation and production of biogas.

the biomass categories and biomass supply ranges as a result of various approaches and methods used by different studies. The major reason for the differences is that the two most crucial parameters – land availability and yield levels – are uncertain, and subject to widely different opinions (e.g. the estimates for 2050 plantation supply ranges from less than 50 EJ yr^{-1} to almost 240 EJ yr^{-1}). In addition, the expectations about future availability of forest wood and of residues from agriculture and forestry vary substantially among the studies.

In theory, with projected technological progress and without jeopardising the world's food supply, energy farming on current agricultural land could contribute over 800 EJ. Organic waste and residues could possibly supply another 40–170 EJ, with uncertain contributions from forest residues and potentially a very significant role for organic waste, especially when bio-materials are used on a larger scale[1]. In total, the upper limit of the bioenergy potential could be over 1,000 EJ annually. This is considerably more than the current global energy use of about 430 EJ.

---

[1] The range of the land area required to meet the potential additional global demand for bio-materials (such as bio-plastics or construction materials) was not included in Table 1. The energy supply of bio-materials ending up as waste can vary between 20-55 EJ (or 1,100-2,900 Mt dry matter) per year. This range excludes cascading and does not take into account the time delay between production of the material and 'release' as (organic) waste.

How do these bottom-up potentials compare to top-down calculations on how much biomass could be produced? In the 1980s and 1990s, the late Prof. DO Hall (at the time the world's leading expert on photosynthesis), and others showed that man already appropriates c. 10% of the global net primary production (NPP) of biomass through agriculture and forestry activities. Dukes (2003) takes this analysis further and extrapolates it to say that the energy fixed through photosynthesis into biomass by this 10% appropriation is approximately equal to current global primary energy demand. Hence, by simple extrapolation, mankind would need to appropriate another 10% of the global NPP to meet a 430 EJ demand solely from bioenergy. In this discussion, it is important to point out that global photosynthetic capacity and therefore NPP is not fixed because limiting factors such as plant nutrients, water and pest and diseases can be managed by farmers and foresters[2].

However, the question of how an expanding bioenergy sector would interact with other land uses, such as food production, biodiversity, soil and nature conservation, and carbon sequestration has been insufficiently analysed in these studies. A refined model of interactions between different uses and bioenergy, food and materials production, would facilitate an improved understanding of the prospects for large-scale bioenergy in the future. Recently, these issues were addressed in several studies. One approach is reported in Smeets *et al.* (2004) where bottom-up information was used on land-use, agricultural management systems on a country-by-country basis, projections for demand for food and information on possible improvements in agricultural management (both for crops and production of meat and dairy products). Figure 1 shows the possible variation in the technical potential, assuming four different agricultural production systems (see also Table 2). In all scenarios, no food shortages occur. Scenarios 1 to 3 have in common that they are based on medium growth assumptions between 1998 and 2050 for global human population (from 5.9 to 8.8 billion people) and per capita food consumption (from 2.8 to 3.2 Mcal per person day), a high plantation establishment scenario (from 123 to 284 Mha) and a high technological level for the production of bioenergy crops. Scenario 4 is based on the assumption that research and development efforts may increase yields above the existing level of technology used in this study as e.g. in scenario 3. In scenario 4 crop yields are 25% higher than in scenario 3 due to additional technological improvements. For further details, see Smeets *et al.* (2004).

---

[2] For example, leaving aside (at this point) the wider environmental and social implications, well-managed sugarcane grown in Brazil's cerrados fixes between 20 and 30 oven dry tonnes of biomass per ha/yr compared to undisturbed 'natural' vegetation which could fix between 0 and 5 odt/ha/.yr when mature. Land management is a crucial factor therefore.

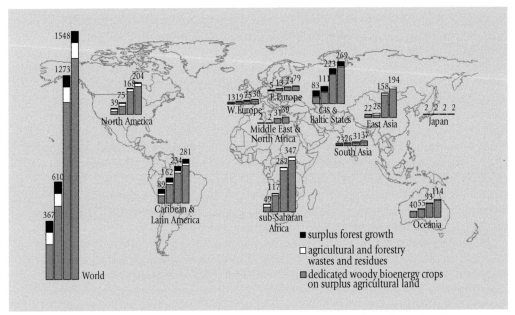

Figure 1. Total bioenergy production potential in 2050, agricultural production systems scenarios 1 to 4. The numbers above the bars are EJ/yr. For more background information, see Table 2 and Smeets et al. (2004).

Other studies carried out by Hoogwijk (Hoogwijk *et al.*, 2005; Hoogwijk, 2004) used integrated assessment modelling to evaluate future biomass potentials for different SRES scenario's. In these analyses, Latin America, Sub-Saharan Africa and Eastern Europe are the most promising regions; Oceania and East and NE Asia also show significant potential in biomass production areas under some scenarios in the longer term. The latter can be explained by the projected demographic developments (possibly declining population in China after 2030) and fast technological progress in agriculture, leading to substantial productivity increases. These analyses also show that a large part of the technical potential for biomass production may be developed at low production costs of about 2 U$/GJ (Hoogwijk, 2004; Rogner *et al.*, 2000).

Technology development

While the main efficiency gains are to be found in agricultural productivity, technological developments can dramatically also improve competitiveness and the efficiency of bioenergy. These gains encompass two major components: conversion of primary biomass to final energy carriers and long distance biomass supply chains (i.e. intercontinental

Table 2. *Overview of assumptions for Agricultural production system scenarios used by Smeets (2004).*

Agricultural production system scenarios	1	2	3	4
Animal production system used (pastoral, mixed, landless)	Mixed	Mixed	Landless	Landless
Feed conversion efficiency	High	High	High	High
Level of technology for crop production[a]	Very high	Very high	Very high	Super-high
Water supply for agriculture	Rain-fed	Irrigated	Irrigated	Irrigated

[a]Scenarios are based on a plausible combination of technologies e.g. a scenario based on a high level of technology for the production of food crops and a low level of technology used in the animal production system (low feed conversion efficiencies) is considered illogical. Going from scenario 1 to 4, the efficiency of food production (expressed in hectares cropland required to meet the projected increase in consumption) increases, thus the area agricultural land claimed for food production decreases.

transport of biomass derived energy carriers) (Hamelinck *et al.*, 2005). Regarding the first component, current production of biofuels for transport is inefficient, from the perspective of the energy balance and the production per hectare. With the exception of ethanol from sugar cane, 'first generation' biofuels, such as ethanol from corn, sugar beet or wheat, or biodiesel from oil seed crops such as rape seed, typically only reach 20%–50% well-to-wheel GHG emissions reductions compared to gasoline (for ethanol) and diesel (for biodiesel) (IEA, 2004)[3]. Also, such schemes are fairly inefficient on a GJ/ha basis, and far from competitive, even with current oil prices. However, there are several 'second generation' technologies in the pipeline, such as ethanol production from lignocellulosic feedstocks, and production of biodiesel using the Fischer-Tropsch process.

---

[3] Note that Brazilian ethanol from sugar cane is the only biofuel currently commercially available, which achieves much higher GHG emission reductions, i.e. 80-90% (IEA, 2004). Also other current biofuels from crops in tropical regions (e.g. biodiesel from jatropha, palm oil etc.) perform better than biofuels from crops grown in temperate regions).

These technologies can achieve higher GHG reduction rates and higher yields per hectare. Also, they will be able to convert a larger diversity of biomass feedstocks than the current first generation technologies, in particular, low cost residucs. It is expected that these second generation technologies will be commercially available within the next one or two decades, i.e. in the time frame in which truly large-scale production of biofuels could take off. These aspects have to be taken into account when calculating future land-requirements. Thus, we deem calculations, such as how much the EU's land area would be needed to cover its domestic demand for e.g. biodiesel based on current rape seed yields and conversion technologies, rather misleading. Regarding the second component of bioenergy logistics, development technologies which convert low-density (both in terms of mass and energy per volume) primary biomass to high-density, high value energy carriers such as wood pellets, torrefied pellets, pyrolysis oil or even directly produced transportation fuels such as ethanol and biodiesel, will widen the possibilities of long-distance bioenergy trade and increase the competitiveness of biofuels.

Perennial crops: the way forward

Regarding the feedstock production, second-generation technologies will favour the production of perennial crops (such as eucalyptus, poplar, grasses such as miscanthus and sugar cane), as they are better than the current annual agricultural crops, economically and environmentally. Next to their better GHG performance, soil carbon improvements can be realised, while fertiliser and pesticide inputs are generally lower. In addition, a recently published article in Nature shows that actual biomass yields can be higher if a large biodiversity of perennial crops is maintained (Tilman, 2006a,b). If designed and managed wisely, biomass plantations can be multi-functional and may generate local environmental benefits. For example, willow plantations in Sweden may be used for soil carbon accumulation, increased soil fertility, reduced nutrient leaching, shelter belts for the prevention of soil erosion, plantations for the removal of cadmium from contaminated arable land (phyto-remediation), and vegetation filters for the treatment of nutrient-rich, polluted water (Börjesson and Berndes, 2006).

Short rotation woody crops (SRC) in general require fewer inputs of herbicides and pesticides. Rich *et al.* (2001) suggest SRC plantations are generally better for a wide variety of wildlife than existing adjacent farmland around the (former) ARBRE project area (UK). When established on agricultural land an increase in biodiversity usually result, e.g. in some cases an increase in species richness occurs. SRC is generally regarded as environmentally friendly and many environmental groups view the technology favourably. Also, in the UK, large scale SRC monoculture is unlikely given the nature of land tenure. Rather, the most likely scenario may be a large number of small plots scattered over large areas.

What are the main critical issues regarding the large-scale production of biomass for energy?

The (sustainable) use of different types of land (marginal and degraded, as well as good quality agricultural and pasture land) depends on the success of accelerating the improvements in current agricultural management practices, and integrating biomass production in a sustainable way into current land-use patterns. Our understanding of how this can be achieved from region to region is often limited. Current experiences with energy crops such as willow, miscanthus and switchgrass, is limited but can point to how biomass production can gradually be introduced in agriculture and forestry. In developing countries (e.g. in sub-Saharan Africa) very large improvements can be made in agricultural productivity[4]. However, better and more efficient agricultural methods cannot not be implemented without investments, proper capacity building and infrastructure improvements and political stability. Much more experience is needed with such schemes, in which the introduction of bioenergy can play a pivotal role to create more income for rural regions by additional bioenergy production. Financial resources generated could then accelerate investment in conventional agriculture and infrastructure and also lead to improved management of agricultural land. Critical issues that require further research and especially more regional demonstrations and experience with biomass production are:

1. Competition for water

Water is a critical resource for both food and biomass production and in short supply in many regions. Water scarcity in relation to additional biomass production has been addressed to a limited extent. Berndes (2002) explains that 'large-scale expansion of energy crop production would lead to a large increase in evapotranspiration appropriation for human uses, potentially as large as the present evapotranspiration from global cropland. In some countries this could exacerbate an already stressed water situation. But there are other countries where such impacts are less likely to occur. One major conclusion for future research is that assessments of bioenergy potentials need to consider restrictions

---

[4] Current agricultural methods deployed in sub-Saharan Africa are is often subsistence farming, with low yields per hectare.

from competing demand for water resources. Improved agricultural must enhance water-use efficiency (e.g. through breeding for drought tolerance and by using drip instead of overhead irrigation).

2. Availability of fertilisers and pest control

Raising agricultural productivity can only be achieved when better management and higher productivities are achieved. This implies better plant nutrition and pest control methods. Sound agricultural methods (agroforestry, precision farming, biological pest control, etc.) exist that can achieve major increases in productivity with neutral or even positive environmental impacts. However, such practices must be secured by sufficient knowledge, funds and human capacity.

3. Land-use planning taking biodiversity and soil quality into account

Criticism is raised by various recent studies (e.g. by the MNP (ten Brink *et al.*, 2006) and the European Environment Agency (EEA, 2006)) that further intensification of agriculture and large scale production of biomass energy crops may result in a losses of biodiversity compared to current land-use, even when international standards for nature protection (10-20% of land reserved for nature) are respected (Smeets *et al.*, 2004). Perennial crops have a better ecological profile than annual crops and the benefits with respect to biodiversity can be achieved when perennial crops are displaced. However, insights into how biodiversity can be optimised and improved compared to current land-use, when sound landscape planning is introduced, are limited and urgently require R&D. Overall, experience from Sweden and the UK with integration of willow production on landscape level, indicate there is a positive effect. Sao Paulo State has strict standards for sugar cane production areas and which appear to ensure that its production does not necessarily lead to a loss in biodiversity. More regional efforts, experience and site-specific solutions are needed. Regarding improvement of soil quality, Lal (2006) shows how some biofuel plantations e.g. Jatropha, Pongamia, can contribute to restore degraded soil and sequester carbon in biota and soil.

4. The use and conversion of pasture land

As discussed above, much land can be released when production of meat and dairy products is done in more intensively (including partial zero-grazing). This would allow grassland currently used as pasture to be used more efficiently. Grasslands could then be used for production of energy grasses or partly be converted to woodlands. Such changes in land-use functions have been poorly studied. The impacts of such changes should be closely evaluated.

5. Socio-economic impacts

Large scale production of modern biofuels, could provide a major opportunity for many rural regions around the world to generate income and employment. Given the size of the global market for transport fuels, the benefits could be vast, e.g. by reducing oil imports and exporting biofuels. Nevertheless, it is far from certain that those benefits will accrue to the rural populations and small-holder farmers. Also, the net impacts for a region as whole, including possible changes and improvements in agricultural production methods, should be kept in mind when developing biomass and biofuel production capacity. New biofuel production schemes should ensure the involvement of the regional stakeholders, in particular the farmers. Worldwide experience with such schemes needs to be developed.

6. Macro-economic impacts of changes in land-use patterns

Although the analyses discussed indicate that both world food demand and additional biomass production *can* be reconciled, more intensive / efficient land-use and additional land-use for biomass production may lead to macro-economic effects on land and food prices. Although this is not necessarily a bad outcome as it could be vital for farmers to enable investment in current production methods, the possible implications on macro-economic level are poorly understood. Again, more work is needed to identify the speed at which changes should be implemented to avoid undesired economic effects[5].

7. Net GHG emissions – including indirect land-use effects

Connected to the previous issue are impacts on overall GHG emission rates related to (biomass-induced and general) changes in land-use. As pointed out previously by Cameron (2006), the pressure on land is often huge in many developing countries such as Malaysia and Indonesia, and to a lesser extent Brazil. In these cases, increasing production of palm oil and soya are one of the main drivers of deforestation, and GHG emissions arising from forest clearance by fire, and changing soil carbon stocks with different types of land use. If biomass energy crops increase pressure on land, these problems could be exacerbated, both directly

---

[5] For a more detailed treatment of the biomass vs. fuel debate, see for example the SEI Newsletter [June 2005].

and indirectly. For the direct cases, more research is required on GHG balances when perennial energy crops replace pastures, (degraded) farm land or forests – the choice of the right cropping system is crucial. Regarding the induced impacts, it is clear that land-use change patterns are complex, and that whole-system GHG emissions have to be assessed.

Tackling the issues: development criteria for sustainable biomass production

With the increasing international trade in biomass resources, concerns have been growing about whether all imported biomass streams can be considered sustainable. The production and removal of biomass can have negative impacts on ecology and land-use, as well as socio-economic impacts and GHG emissions. Recently, these aspects have been recognised by policy makers, scientists and the industry. Various preliminary efforts have been undertaken to move towards certification and track-and-trace systems for imported biomass. Examples include the development of the Green-Gold-Label, a biomass tracking system developed by Essent (CU, 2006), the FairBiotrade research project carried out by Copernicus UU (see e.g. Lewandowski and Faaij, 2006, Damen and Faaij, 2006; Smeets and Faaij, 2007), and various other studies on sustainability and certification of biomass (see e.g. Tipper *et al.*, 2006; WWI, 2006; WWF, 2006). Furthermore, the initiatives such as the IEA Bioenergy Task 40 on International Sustainable Bioenergy Trade (see www.bioenergtrade.org), the FAO International Bioenergy platform (IBEP) or the UNCTAD Biofuels initiative demonstrate the increased attention to global biomass trade and sustainability.

The need for biomass sustainability criteria has also been recognised in several EU countries and by different international bodies. Current examples are:
- Ongoing development of GHG and sustainability criteria for biomass transportation fuels under the renewable transport fuel obligation (RTFO) in the UK (Archer, 2006).
- Existing regulations energy/CO_2 balances and sustainability criteria for Belgian biomass for co-firing (Ryckmans, 2006).
- The EU strategy for biofuels (EC, 2006), in which standards to ensure the sustainability of biofuel feedstocks are explicitly mentioned.
- More in general, the issues surrounding the production of palm oil in Southeast Asia and Soy beans in South America have triggered the establishment of round tables where all stakeholders in the chain are represented.

The Dutch government has one of the most advanced policies for developing sustainability criteria for biomass. In the autumn of 2005, awareness regarding the necessity of biomass sustainability criteria increased when environmental NGOs condemned the use of palm oil for green electricity production in natural gas-fired power plants. While the short-term policy reaction was to reduce feed-in tariffs for palm oil, the urgent need for biomass sustainability criteria was recognised by the Dutch parliament. Thus, a commission was established in January 2006 to develop a system for biomass sustainability criteria for the Netherlands.

The main starting points of the commission were (Cramer *et al.*, 2006):
- Development of a long-term vision about biomass sustainability (2020-2040).
- Based on this vision, development of concrete, measurable biomass sustainability criteria in the short term.
- Development of a universal framework of sustainability criteria, with the emphasis on non-food applications (chemical industry, fuels, energy production). The sustainability criteria and indicators developed could also be of importance to judge food production on sustainability aspects. It is acknowledged that biomass, feed, fuel and fodder can barely be regarded separately.
- Compliance with international treaties, EU regulations, WTO rules, etc.
- Development of short term minimum sustainability demands and stricter criteria in the longer term.
- Sustainability criteria are valid for both biomass energy crops and biomass crops, and both applicable for imported biomass and domestic biomass.

Based on these starting points, consultations with Dutch stakeholders and scientific support, the commission developed a number of biomass sustainability criteria and indicators/procedures for the short-term (2007) and the medium term (2011). These included GHG reductions of at least 30% (rising to 50% by 2011), no decline of biodiversity or valuable ecosystems, prevention of soil erosion, preservation of quality and quantity of surface water and ground water increased human welfare and no reduction in food supplies, etc. (see Table 3 for more details; Cramer *et al.*, 2006).

While it is clear that for most of such criteria, indicators and procedures still need to be developed, these approaches show promise. What is more important to emphasise is that such criteria cannot be developed overnight. The procedure is to set minimum levels of sustainability criteria now, and use pilot cases to build up experience of how sustainability criteria can be met under diverse conditions. Also, the proposed sustainability goes far beyond many other sectors. This

Table 3. Summary of sustainability criteria, indicators/procedures and suggested levels for 2007 and 2011 (Cramer et al., 2006). For criteria 2-6 a dialog with national and local stakeholders is required.

Criterion and level	Indicator/procedure 2007	2011
1. GHG balance Net emission reduction by >=30% in 2007 and >= 50% in 2011	Use of developed methodology Use of reference values for specific steps in logistic chain	As 2007
2. Competition with food supply, local energy supply, medicines and building materials Supply is not allowed to decrease	Footnote a	Footnote b
3. Biodiversity No decline of protected areas or valuable ecosystems in 2007, also active protection of local eco-systems in 2011	No plantations near gazetted protected areas or high conservation value areas Max. 5% conversion of forest to plantations within 5 years Footnote a	As 2007 Additional obligatory management plan for active protection of local ecosystems Footnote b
4. Wealth No negative effects on regional and national economy in 2007, and active contribution to increase of local wealth in 2011	Footnote a, based on Economic Performance indicators of the global reporting initiative	Footnote b
5. Welfare, including a. Labour conditions	Compliance with Social Acountability 8000 and other treaties	As 2007
b. Human rights	Compliance with universal declaration of HR	As 2007
c. Property and use rights	Three criteria from existing systems (RSPO 2.3, FSC 2, FSC 3)	As 2007
d. Social conditions of local population	Footnote a	Footnote b
e. Integrity	Compliance with Business principles of countering bribery	As 2007

Table 3. Continued.

Criterion and level	Indicator/procedure 2007	2011
6. Environment, including		
a. Waste management	Compliance with local & national laws GAP	As 2007
b. Use of agro-chemicals (incl. fertilisers)	Compliance with local & national laws	As 2007 & EU legislation
c. Prevention of soil erosion and nutrient depletion	Erosion management plan Avoid plantations on marginal or vulnerable soils, or with high declivity Monitoring soil quality Nutrient balance	Footnote b
d. Preservation of quality and quantity of surface water and ground water	Footnote a, special attention for water use and treatment	Footnote b
e. Airborne emissions	Comply with national laws	Comply with EU laws
f. Use of GMOs	Compliance with USA (safety) rules	Compliance with European (safety) rules

[a]For this criterion a reporting obligation applies. A protocol for reporting will be developed.
[b]New performance indicators will be developed for this criterion between 2007-2011.

could easily backfire on biotrade if too many restrictions are put in place, making an evaluation period even more important. In addition, some sustainability criteria may actually conflict with each other[6] and, the costs of meeting the sustainability criteria will have to be evaluated[7] and if necessary the criteria and indicators can be adapted and improved. This was the approach followed in the Dutch case, and a four-year evaluation period has been established.

---

[6] For example modernisation of agriculture may make the necessity of very hard physical labour obsolete. At the same time greater mechanisation will lead to less employment.

[7] Examples of other certification system show that depending on the local situation and specific criteria, additional costs may vary widely, e.g. 8-65% (Van Dam *et al.*, 2007).

Finally, a crucial aspect of such criteria is enforcement. Examples from FSC-certified wood show that such systems are effective but also not flawless. Frequent field visits are vital to ensure compliance with criteria, as is stakeholder participation both during the set-up and monitoring of certification systems.

The role of international bioenergy trade

After discussing the potentials and pitfalls of global bioenergy production we want to emphasise, be it briefly, the importance of international bioenergy trade as one of the main drivers behind development of the major transitions required. As can be seen e.g. in Figure 1, main supply regions are generally not situated in densely populated and highly developed areas where demand is/can be large. Bioenergy trade is developing rapidly and is already proving to be substantial as highlighted through these three brief examples:

Pellet exports from Canada to European countries and the USA

The export of wood pellets from Canada has grown exponentially in the past several years (see Figure 2), primarily from the west coast. There are at least 11 pellet plants in Canada, exporting to Europe and the US (Bradley, 2006). Expectations are that production will exceed one million tonnes in 2006.

Ethanol exports from Brazil to Japan, the USA and Europe

Figure 3 shows Brazil's ethanol trade since 1970. Market opportunities and constraints have determined exports and imports. A substantial amount of ethanol was imported during the 1990s, first during the supply shortage of ethanol (1990-1991) and second when international sugar markets were favourable for exports (1993-1997). Traditionally, Brazilian exports of ethanol have been oriented for beverage production and industrial purposes but, recently, trade for fuel purposes has increased significantly, as illustrated in Figure 3,. In 2004 exports reached 2.5 billion litres and it is estimated that almost the same amount was exported in 2005 (Walter et al., 2006).

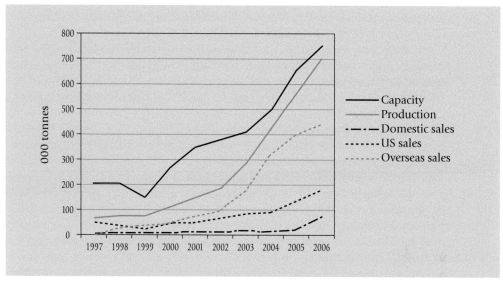

Figure 2. Pellet production and export in Western Canada (Bradley, 2006).

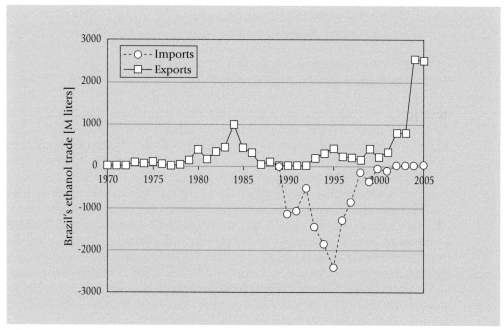

Figure 3. Trade in ethanol in Brazil 1970-2005 (Walter et al., 2006).

Palm oil and palm kernel shell exports from Malaysia and Indonesia to Europe

Over the last years, increasing amounts of palm kernel shells and palm oil have been co-fired in European power plants. While no exact statistics are available, substantial imports have been occurring in the UK and the Netherlands (Junginger *et al.*, 2006; Rosillo-Calle and Perry, 2006).

These examples show how international bioenergy trade helps to cover the demand for transport biofuels and for electricity from biomass. The future vision on global bioenergy trade is that it develops over time into a real 'commodity market' which will secure supply and demand in a sustainable way. The development of truly international markets for biomass may become an essential driver to deliver the biomass potentials discussed above and exploit a resource which is currently under-utilised in many world regions. Exporting biomass-derived commodities to supply the world's energy markets could provide a stable and reliable demand for rural communities in many countries, particularly developing ones, thus creating an important incentive for rural investment that is much needed in many areas in the world. Thus, we see trade as an essential prerequisite for viable bioenergy development, with the practical monitoring of sustainability as a key factor for long-term security.

How much energy can biomass deliver?

The techno-economic potential of biomass resources for energy and industrial materials can be very large.I In theory, twice the current global energy demand, but more likely around 400 EJ, without competing with food production, protection of forests and nature. Roughly, one quarter (100 EJ) could be provided by efficiently exploiting residues from agriculture and forestry and from organic waste. Another 100 EJ could stem from the rehabilitation of degraded land. Note that these two potentials do not require additional land. The remaining half could come from dedicated energy crops on current agricultural and pasture lands, corresponding to about 1 billion ha worldwide. This is some 8% of the global land surface and one-fifth of the land currently in use for agricultural production. If the global bioenergy market is to develop to supply 400 EJ per year over this century (compared to 430 EJ current total global energy use), the value of that market assuming US$4/GJ would amount to some $1.6 trillion per year. Logically, not all biomass will be traded on international markets, but such an indicative estimate how important this market could become for rural areas worldwide.

These numbers are impressive, perhaps daunting to some. Major transitions are required to exploit this bioenergy potential, which can only be reached in the second half of this century. Improving agricultural land-use efficiency in developing countries (i.e. increasing crop yields per hectare) is a key factor. It is still uncertain to what extent and how fast such transitions can be realised in different regions. Significant problems are posed by the lack of capital, skills, land tenure, etc., all of which are major impediments to agricultural modernisation. Under less favourable conditions, the (regional) bioenergy potential(s) could be quite low.

Biofuels are not the panacea for solving the global energy situation, but should be seen as part of the solution. While bioenergy can, in theory, provide a substantial part to the future global energy supply, realising these potentials will require profound changes, especially in agriculture. There are potential alternatives to bioenergy that can also play a major and sometimes synergistic role e.g. solar, geothermal, wind, etc. Even though there are a number of critical issues involved with the large-scale production of biofuels, we also see opportunities for many countries e.g. restoring degraded soils using biomass could result in environmental gains and exporting refined biofuels can be high-value export products allowing re-investment in poor rural areas.

Ensuring the sustainability of biomass production is a major challenge, but also a great opportunity. Change means not only threats, but also opportunities. The challenge is to be able to implement the change in intelligent and benign ways. After all, this could be a first large-scale commodity market where there is a considerable scope for implementing sustainability criteria – which, in turn, could have positive impacts on food and fodder commodities. At the same time, global bioenergy trade is growing rapidly, and annual increases of 100% of traded biomass volumes are becoming reality. Therefore, the rapid early development and implementation of sustainability frameworks is crucial.

Certification, preferably starting from an internationally accepted framework but applied and verified at a regional level with strong stakeholder participation, seems to be a feasible way to achieve this. Showing best-practice operations through export-oriented pilot projects in a diversity of developing countries and different rural areas is crucial in the short term. Good examples of successful business models and sound sustainability frameworks can guide market forces in a sustainable direction. If we succeed, we may be looking at the first stages of the Green OPEC (or BIO-PEC) of the future!

References

Archer, G., 2006. Development of Carbon Certification & Sustainability Assurance for Biofuels in the UK. Presentation held at A Sustainable Path for Biofuels in the EU, 7th June 2006 Brussels, Belgium.

Berndes, G., 2002. Bioenergy and water—the implications of large-scale bioenergy production for water use and supply. Global Environmental Change 12:253-271.

Berndes, G., Hoogwijk, M. and van den Broek, R., 2003. The contribution of biomass in the future global energy supply: a review of 17 studies. Biomass and Bioenergy 25:1-28.

Börjesson, P. and G. Berndes, 2006, The prospects for willow plantations for wastewater treatment in Sweden. Biomass and Bioenergy 30:428-438.

Bradley, D., 2006. Canada Biomass-Bioenergy Report. Canadian country report for IEA Bioenergy Task 40, available at www.bioenergtrade.org, May 2006.

CU, 2006. Green Gold Label. Certification system for sustainable biomass. Control union, available at www.controlunion.com/certification/program/Program.aspx?Program_ID=19

Cameron, A., 2006. Green or grey? Sustainability issues of biofuel production. Renewable Energy World 02/03/06, available at http://www.earthscan.co.uk/news/article/mps/uan/638/v/3/sp/

Cramer, J. (lead author), 2006, Criteria voor duurzame biomassa productie. Final report of the project group 'Duurzame productie van biomassa', 14 July 2006, 33 pp.

Damen, K. and A. Faaij, 2006. A Greenhouse gas balance of two existing international biomass import chains; the case of residue co-firing in a pulverised coal-fired power plant in the Netherlands. Mitigation and Adaptation Strategies for Global Change (Special Issue) 11: 1023-1050.

Dukes, J.S., 2003. Burning buried sunshine: human consumption of ancient solar energy. Climatic Change 61: 31-44.

EEA, 2006. How much bioenergy can Europe produce without harming the environment? European Environment Agency, EEA report 7/2006, p. 67.

EC, 2006. An EU strategy for biofuels. Communication from the commission, {SEC(2006) 142}, Brussels, 8.2.2006, p. 29.

Faaij, A., van den Broek, R., Lysen, E., Gielen, D., Hoogwijk, M. and Wolf, J., 2000. Mondiale beschikbaarheid en mogelijkheden voor import van biomassa voor energie in Nederland: Synthese van het onderzoeksproject GRAIN: Global Restrictions on biomass Availability for Import to the Netherlands. Vakgroep Natuurwetenschap & Samenleving - Universiteit Utrecht, UCE, ECN, LUW-TPE. Rapportage in opdracht van NOVEM, Utrecht, Juli 2000.

Hamelinck, C.N., R.A.A. Suurs and A.P.C. Faaij, 2005. Techno-economic analysis of International Bio-energy Trade Chains. Biomass & Bioenergy 29: 114-134.

Hoogwijk, M., 2004. On the global and regional potential of renewable energy sources. PhD thesis, Copernicus Institute, Utrecht University, 256 pp.

Hoogwijk, M., Faaij, A., van den Broek, R., Berndes, G., Gielen, D. and Turkenburg, W., 2003. Exploration of the ranges of the global potential of biomass for energy. Biomass and Bioenergy 25:119-133.

Hoogwijk, M., Faaij, A., Eickhout, B., de Vries, B. and Turkenburg, W., 2005. Potential of biomass energy out to 2100, for four IPCC SRES land-use scenarios. Biomass & Bioenergy 29:225-257.

IEA, 2004. Biofuels for transport. An international perspective. International Energy Agency, Paris, France.

Junginger, M., M. De Wit and A. Faaij, 2006. Country report for IEA Bioenergy Task 40, update 2005, draft.

Lal, R., 2006. Land area for establishing biofuel plantations. Energy for sustainable development 10:67-79.

Lewandowski, I. and Faaij, A., 2006. Steps towards the development of a certification system for sustainable bioenergy trade. Biomass & Bioenergy 30:83-104.

Rich, T.J., Sage, R., Moore N., Robertson P., Aegerter J. and Bishop J., 2001. ARBRE Monitoring-Ecology of SRC Plantations (Interim Report), ETSU B/U1/00627/REP; London, UK.

Rogner, H-H., Barthel, F., Cabrera, M., Faaij, A., Giroux, M., Hall, D., Kagramanian, V., Kononov, S., Lefevre, T., Moreira, R., Nötstaller, R., Odell, P., and Taylor, M., 2000. Energy Resources. Chapter 5 of the World Energy Assessment of the United Nations, UNDP, UNDESA/WEC, Published by: UNDP, New York, September 2000.

Rosillo-Calle, F. and Perry, M., 2006. UK country report for IEA Bioenergy Task 40, draft.

Ryckmans, Y., 2006. Implementation strategy for large scale biomass imports. Presentation given at the T38/T40 workshop in Trondheim, Norway, April 2006, available at www. bioenergy trade.org.

SEI Newsletter, June 2005. Policy Debate on Global Biofuels Development. Available at: http://www.sei.se/red/RED-June-2005.pdf

Smeets, E. and Faaij, A., 2007. Bioenergy potentials from forestry to 2050. Climatic Change 81:353-390.

Smeets, E., Faaij, A., Lewandowski, I., 2004. A quickscan of global bioenergy potentials to 2050 – an analysis of the regional availability of biomass resources for export in relation to underlying factors, Report prepared for NOVEM and Essent, Copernicus Institute – Utrecht University, NWS-E-2004-109, March 2004. Pp. 67 + Appendices.

Smeets, E., Faaij, A. and Lewandowski, I., 2005. An exploration of the impact of the implementation of sustainability criteria on the costs and potential of bioenergy production applied for case studies in Brazil and Ukraine. Report prepared for NOVEM and Essent, Copernicus Institute – Utrecht University, NWS-E-2005-6, ISBN 90-73958-00-8, February 2005. 78 pp. plus appendices.

Ten Brink, B. (Lead author) et al., 2006. Cross-roads of Planet Earth's Life. Exploring means to meet the 2010 biodiversity target. Study performed for the Global Biodiversity Outlook 2 Chapter solution-oriented scenarios. MNP, report 555050001/2006, pp. 92.

Tipper, R., Garstang, J., Vorley.W. and Woods, J. Draft Environmental Standards for Biofuels. Low-Carbon Vehicles Partnership, London. 2006.

Tilman, D., 2006a. Natural prairie holds key to sustainable fuels. UMN news, May 31, 2006, available at: www1.umn.edu/umnnews/Feature_Stories/Natural_prairie_holds_key_to_sustainable_fuels.html.

Tilman, D., Reich, P.B. and Knops, J.M.H., 2006b. Biodiversity and ecosystem stability in a decade-long grassland experiment. Nature 441:629-631.

Van Dam, J., M. Junginger, A. Faaij, I. Jürgens, G. Best and U. Fritsche, 2007. Overview of recent developments in sustainable biomass certification. Biomass & Bioenergy (special issue on international bio-energy trade).

Walter, A., Dolzan, P. and Piacente, E., 2006. Biomass Energy and Bioenergy Trade: Historic Developments in Brazil and Current Opportunities. Brazilian country report for IEA Bioenergy Task 40, available at www.bioenergtrade.org, June 2006.

WWF, 2006. The potential for delivering sustainable bioenergy. Background & discussion paper. Draft for the WWF workshop: The way forward to sustainable bioenergies. European Commission – DG TREN. Brussels, 29 June 2006.

WWI, 2006. Biofuels for transportation. Global potential and implications for sustainable agriculture and energy in the 21st century. Final report. Prepared by the Worldwatch Institute for the German Federal Ministry of Food, Agriculture and Consumer Protection (BMELV), in cooperation with the Agency for Technical Cooperation (GTZ) and the Agency of Renewable Resources (FNR) Washington, D.C. June 7, 2006.

Outlook for U.S. ethanol and DDGS production and disposition

G. Cooper
National Corn Growers Association, 632 Cepi Dr., Chesterfield, Missouri, 63005, USA

Abstract

Rapid expansion of the U.S. ethanol industry is changing the face of world agriculture markets. A strong emphasis on converting agricultural resources into energy has increased demand for grain crops, therefore encouraging farmers to increase production. Corn production projections indicate an adequate supply of corn will be available for export demand in the long-term. However, increased demand is likely to lead to an incremental rise in corn price. Therefore, U.S. livestock and poultry feeders are investigating the prospects of substituting other feedstuffs into their rations to offset expected increases in corn price. A result of the ethanol boom has been an ever-increasing supply of distillers grains, which have been economically utilised in beef, dairy, swine, and poultry rations. Though distillers grains products have experienced consistency and transportation issues in the past, usage continues to grow dramatically in conjunction with supply. Corn distillers grains are high in energy and protein and can be used in many different types of rations. The majority of U.S.-produced distillers grains historically have been consumed domestically, though export demand has shown signs of increasing in recent years. New dry-grind fractionation processes promise to add another level of complexity to a rapidly changing market. As ethanol production continues significant growth in the United States, it behoves domestic and foreign feed buyers and livestock producers to gain a better understanding of the issues associated with purchasing, transporting, and feeding distillers grains.

Keywords: corn, ethanol, distillers grains, distillers dried grains with solubles (DDGS), corn gluten

Introduction

As the world's largest producer of corn, the United States has for decades provided domestic and global markets with a consistent supply of grain for livestock feed and human food use. In recent years, however, the U.S. agriculture sector has intensified its focus on developing another end use for corn - energy. Rising world crude oil prices and other factors have stimulated a significant increase in the amount of U.S. corn used for the production of ethyl alcohol, a liquid transportation fuel.

The amount of corn used for ethanol has grown 316% in the past 10 years (Figure 1). In the 1995-1996 marketing year[8], the U.S. ethanol industry consumed 10.1 million metric tons (mmt) of corn, accounting for 4.6% of total corn usage. By comparison, the ethanol industry processed 41.9 mmt in the 2005-2006 marketing year, representing 14.6% percent of total use. It is estimated that the ethanol industry will require 61.98 mmt in 2006-2007, meaning ethanol would surpass exports as the number two use of U.S. corn for the first time in history.

As of September 2006, there were 105 ethanol plants in operation in the United States with total capacity of 18.96 billion liters per year (bly). Additionally, 41 new plants were under construction and seven existing plants were expanding. The combined annual capacity of these new plants and expansions is 11.09 bl (RFA, 2006). At least 110 additional ethanol plants are being proposed, though it is unlikely all of these plants will be built (Figure 2). It is projected that the ethanol industry will be producing between 44-50 bl by the 2011-2012 marketing year.

The growth in ethanol production has been accompanied by an increasing supply of the residual grains that result from the ethanol fermentation process. These residual grains are used as livestock and poultry feed. Today, approximately 30% percent of fuel ethanol is manufactured by wet mills, which produce corn gluten feed (wet or dried), corn gluten meal, and corn germ meal as primary co-products.

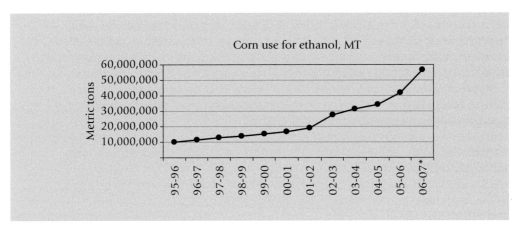

Figure 1. U.S. corn use for fuel ethanol production (USDA, 2006).
*Year 2006-2007, projected.

[8] The USDA marketing year is defined as Sept. 1 through Aug. 31.

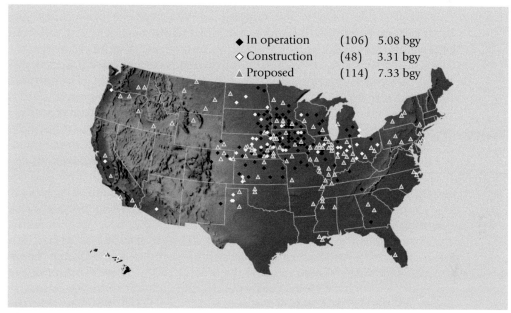

Figure 2. U.S. ethanol plants operating, under construction, and proposed by Sept. 2006 (Renewable Fuel Association, 2006)

The remaining 70% percent of fuel ethanol is produced by dry-grind plants, which generate distillers wet or dried grains (DWG, DDG), distillers wet or dried grains with solubles (DWGS, DDGS), modified distillers wet grains (MDWG), and condensed distillers solubles (CDS) as co-products. These dry-grind ethanol co-products are collectively referred to as distillers grains. This paper will focus largely on DDGS because this is the dry-grind ethanol co-product most commonly exported to international markets.

Several precipitating market and political factors are responsible for the feverish pace of U.S. ethanol plant construction in the past 18 months. First, historically high prices for oil and gasoline in recent years have made the economics of ethanol production considerably attractive. When the federal ethanol income tax credit is taken into account, ethanol production is cost competitive with gasoline when crude oil is $25 per barrel (McClaren, 2006). West Texas Intermediate (WTI) crude oil prices averaged $65.56 per barrel between July 2005 and July 2006 (DOE, 2006a). WTI crude oil prices reached a record high monthly average price of $74.41 per barrel in July 2006. In response to high oil prices, retail prices for regular unleaded gasoline averaged $.69 per liter between July 1, 2005, and Sept. 18, 2006. This compares to an average price of $.51 for the period of Jan. 1, 2004, to June 30, 2005 (DOE, 2006b).

Second, the federal government in the past several years has demonstrated significant support for renewable fuels through enactment of legislation that encourages production and use of renewable energy. The Energy Policy Act that passed in August 2005 included a Renewable Fuels Standard that requires the use of at least 28.4 bl of renewable fuels annually by 2012. Ethanol will constitute the majority of fuel used to meet this requirement, with biodiesel accounting for a much smaller share. The legislation effectively guarantees a market for a substantial amount of ethanol. Additionally, the Congress voted in 2004 to extend a 51-cent-per-gallon income tax credit for ethanol until 2010.

Third, ethanol demand and prices reached unprecedented levels in the spring and summer of 2006 as the U.S. fuels market rushed to replace the gasoline additive methyl tertiary butyl ether (MTBE) with ethanol. For many years, MTBE was ethanol's main competitor as a gasoline oxygenate. In the late 1990s, however, it was determined that MTBE had contaminated groundwater in several states and was a possible threat to human health (Yacobucci and Womach, 2003). Gasoline marketers voluntary chose to cease blending MTBE into gasoline after the Congress opted not to include liability protections for MTBE producers in the Energy Policy Act of 2005. The rapid phase-out of MTBE in 2006 caused a pronounced spike in ethanol demand, and in turn, ethanol prices.

Finally, corn prices remained at relatively low levels in 2005-2006, contributing to the record profit margins experienced in 2006. Ethanol's cost of production is largely dependant on the price of corn, as corn is the single most important input in the ethanol process. The national average farm price for corn in the 2005-2006 marketing year was $77.61 per metric ton, $11.31 below the 10-year average (USDA, 2006). In many top ethanol producing states, the farm price for corn in 2005-2006 was well below the national average.

Growth in the production of distillers grains

The attractive economics of ethanol in the last several years led to an influx of capital and extraordinary growth in the industry. All of the ethanol plants built in the past two years have utilised the dry-grind process. Accordingly, the production of distillers grains has increased dramatically over the past several years (Figure 3).

The production of distillers grains more than doubled from 3.99 million metric tons in the 2001-2002 marketing year to 8.35 million metric tons in 2005-2006. Industry analysts predict distillers grains production levels will reach 12-13 million metric tons in 2006-2007. The remarkable growth in distillers grains production has brought about several important changes in the U.S. feed industry. Traditional diet formulations for beef and dairy have changed, as have swine and

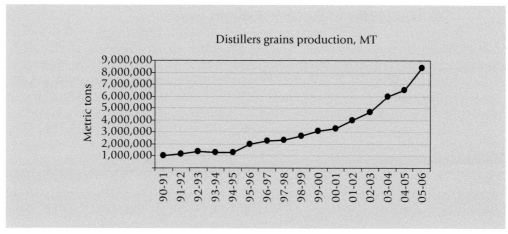

Figure 3. U.S. production of distillers grains (ProExporter Network, 2006).

poultry rations to a lesser extent; the economics of feeding livestock and poultry have changed; and transportation, storage, and handling infrastructure has been adapted. In the near-term, it is expected that international feed markets also will be compelled to continue to adapt to the increasing supply of U.S. distillers grains.

Currently, approximately 35% of distillers grains are marketed domestically as distillers wet grains with solubles (DWGS). Because of its high moisture content (typically 63-68%) and relatively short shelf life, DWGS are normally marketed to beef and dairy operations that are in close proximity to ethanol plants. For example, it is estimated that approximately 70% of the distillers grains produced by dry-grind ethanol plants in Kansas and Nebraska are marketed as DWGS because of the high concentration of fed cattle in those states. Ethanol producers who market their distillers grains locally in the form of DWGS are able to avoid the cost of drying their product, which is often as much as 20% of the plant's total production cost.

The remaining 65% of distillers grains are dried and mixed with solubles to form distillers dried grains with solubles (DDGS). DDGS are relatively low in moisture content (typically 10-12%), making the product suitable for export and long-haul domestic shipments.

Nutrient profile of modern distillers grains

Every bushel of corn entering the dry-grind ethanol process yields nearly equal parts ethanol, distillers grains, and carbon dioxide[9]. Therefore, the nutrient content of corn DDGS can be approximated by multiplying the nutrients in corn by a factor of three. Table 1 shows the averages and ranges of select nutrients from 32 different U.S. corn DDGS sources.

Table 1. Average select nutrient content of 32 U.S. corn DDGS samples (Shurson and Noll, 2005).

Nutrient	Average %	Range
Dry matter	89.3	87.3 - 92.4
Crude protein	30.9	28.7 - 32.9
Crude fat	10.7	8.8 - 12.4
Crude fiber	7.2	5.4 - 10.4
Ash	6.0	3.0 - 9.8
Lysine	0.90	0.61 - 1.06
Phosphorous	0.75	0.42 - 0.99

Product variation and its causes

As demonstrated by the ranges in Table 1, the nutrient content of corn DDGS can vary widely from source to source. The nutrient profile of DDGS from the same source may even vary over time. Numerous factors are responsible for the variation in nutrient content and physical characteristics of DDGS, though three major production factors appear to be the chief causes of inconsistent product.

First, variation in the nutrient content of incoming corn is likely responsible for variance in the finished DDGS. The Nebraska Corn Board's 2005-2006 Corn Quality Report showed that the crude protein content of corn produced in Nebraska ranged from 7.6 to 11.1%, oil content ranged from 3.5 to 5%, and starch content ranged from 70.4 to 73.8% (NCB, 2006). Previous studies also showed wide ranges in lysine and phosphorous content (see Reese and Lewis, 1989). As described earlier, the nutrient content of corn is concentrated by a factor of three

---

[9] 2.8 gallons of fuel-grade ethanol, 17 pounds of distillers grains, and 17 pounds of carbon dioxide.

in the finished DDGS; accordingly, variability of nutrients in DDGS is wider than in the original corn input.

Second, plant to plant variations in corn-to-ethanol fermentation practices likely will affect the consistency of the final DDGS. Even slight changes in the mash cooking temperature, fermentation time, or distillation time will likely impact the consistency of the residual grains. Additionally, plant to plant process variations may lead to inconsistencies in the physical properties of DDGS, most notably micron size and colours.

Finally, an ethanol plant's drying and finishing practices will affect the nutrient profile and physical qualities of DDGS. The length and temperature of drying is widely believed to affect many physical qualities of DDGS, including colour and flowability. Drying discrepancies are likely responsible for the variance in lysine digestibility of the finished DDGS as well. Dryer temperatures vary widely from plant to plant and have been reported to range from as low as 127 °C to 621 °C (260 °F to 1150 °F; Shurson and Noll, 2005). The amount of time DDGS are allowed to 'cure' before load-out is also considered to be a critical factor in the flowability of the product. It is also important to note that the amount of condensed distillers solubles, or syrup, added back to the grains fraction to produce DDGS varies widely from plant to plant. Table 2, which shows the nutrient content of both the solubles fraction and grains fraction, demonstrates that varying solubles application rates have a considerable effect on the nutrient profile of finished DDGS.

Though not a cause of true physical or nutritional variability, many industry experts believe the current system for defining and analysing distillers grains products may contribute to market confusion and

Table 2. Average nutrient content of DDGS and CDS, 100% dry matter basis (Shurson and Noll, 2005).

Nutrient	Percentage DDGS	Percentage CDS (solubles)
Crude protein	33.5	18.5
Crude fat	9.0	15.7
Crude fiber	9.5	2.5
Ash	3.0	8.4
Calcium	0.04	0.06
Phosphorous	0.54	1.28
Lysine	1.05	0.68
Methionine	0.66	0.27
Threonine	1.27	0.70
Trytophan	0.29	0.20

perceived inconsistency. The definition for DDGS used by the industry today was adopted by the American Association of Feed Control Officials (AAFCO) in 1964 when a relatively minute amount of the product was traded in the feed market. The definition, which is considered to be too general in scope by many in the industry, is: 'Corn Distillers Dried Grains with Solubles is the product obtained after the removal of ethyl alcohol by distillation from the yeast fermentation of a grain or a grain mixture by condensing and drying at least ¾ of the solids of the resultant whole stillage by methods employed in the grain distilling industry'.

Additionally, there are no industry standards for nutrient analysis of DDGS and other distillers grains products. A recent study by South Dakota State University determined that at least 10 test methods are commonly used to determine DDGS moisture content, 11 methods are commonly used for crude protein analysis, and nine methods are commonly used for crude fat analysis (Theix, 2006). It is not surprising that using various test methods on a singular DDGS sample may yield significantly different results, a fact demonstrated by 2005 Land O' Lakes/Purina Feed lab experiment on testing variation (Table 3). In this trial, a singular DDGS sample was split and sent to six independent laboratories for analysis of moisture, crude fat, and crude protein. Due to the fact that each lab was using different methods to analyse the split samples, the results varied widely.

Table 3. DDGS analysis variation due to use of different test methods (Tilstra, Land O' Lakes/Purina Feed, LLC, 2005).

Split sample #	Moisture, %	Fat, %	Protein, %
1	12.69	13.73	26.00
2	10.48	10.01	26.30
3	10.09	10.04	27.30
4	10.64	8.71	26.13
5	13.30	10.15	26.29
6	12.60	9.40	25.00
Range	3.21	5.02	2.30

Determining the value of DDGS

The market value of DDGS varies widely based on a number of important factors. While nutrient concentration is chief among these factors, the value of DDGS also is dependent on the value of nutrients in the feedstuff it is replacing. This often depends on whether DDGS is being fed primarily for protein or primarily for energy (total digestible nutrients [TDN]). Table 4 illustrates the replacement value of DDGS on an energy (total digestible nutrient) equivalency level with corn. The equivalent value of DDGS is compared to soybean meal on a crude protein basis in Table 5. Nutrient availability, physical properties of the product, species being fed, and market prices of competing feedstuffs are other major factors that impact the market value of DDGS.

Table 4. Equivalent Value of DDGS compared to corn on an energy (TDN) basis. Assumptions: corn, 88% DM and 88% TDN; DDGS, 90% DM and 97% TDN.

Corn [$/mt, ($/bushel)]	DDGS ($/mt)
59.03 (1.50)	66.41
62.97 (1.60)	70.85
66.90 (1.70)	75.27
70.85 (1.80)	79.70
74.78 (1.90)	84.13
78.72 (2.00)	88.56
82.65 (2.10)	92.99
86.58 (2.20)	97.41
90.52 (2.30)	101.84
94.45 (2.40)	106.27
98.40 (2.50)	110.70

Table 5. Equivalent Value Of DDGS To Soybean Meal (SBM) On CP Basis. Assumptions: SBM, 89% DM And 48% CP; DDGS, 90% DM And 30% CP.

SBM ($/mt)	DDGS ($/mt)
165.30	104.47
176.32	111.43
187.34	118.40
198.36	125.36
209.38	132.33
220.40	139.29
231.42	146.26
242.44	153.22
253.46	160.20
264.48	167.16
275.50	174.13

Despite the fact that modern DDGS has approximately 120% the energy value (TDN) of corn grain for beef (Klopfenstein, 2003), the market price for DDGS has historically tracked the price of corn. Prices have fluctuated widely in the past five years, ranging from a low monthly average of $62/short ton in July 2001 to a high of $128/short ton in April 2003 (Figure 4).

Numerous studies conducted in the United States have highlighted the attractive economics of feeding DDGS to beef, dairy, and swine. Klopfenstein (2003) outlined the economics of feeding beef heifers and cows a soybean meal-based supplement versus a DDG-based supplement. The two supplements had equal feeding value, but the DDG-based supplement was only 62% the cost of the soybean meal-based supplement[10]. The high escape value of the protein in the DDG-based supplement allows the inclusion of less expensive midds and urea in the supplement.

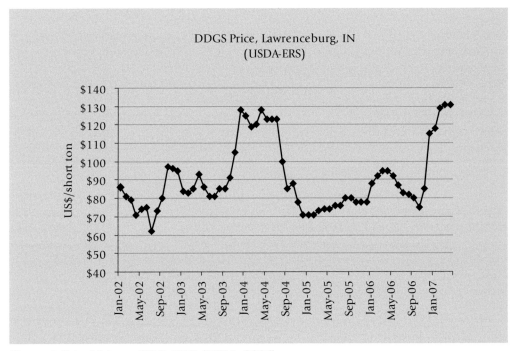

Figure 4. Price history of U.S. DDG (USDA, 2006).

[10] Based on SBM price of $161/ton; DDG, $95; midds, $61; urea, $280; corn, 75$.

Using a simple method outlined by Chase and Linn (1996), the cost per unit of crude protein of DDGS versus soybean meal can be easily determined. For example, if soybean meal is $275.50/mt and DDGS is $104.70/mt, the cost per pound of crude protein in soybean meal is $.616/kg and the cost per pound of crude protein in DDGS is $.374/kg[11].

Thaler (2002) outlined the economics of using 90.7 kg (10 percent of ration) of DDGS and 1.36 kg of limestone to replace 80.8 kg of corn, 8.63 kg of 46% protein soybean meal and 2.72 kg of dicalcium phosphate in a short ton (907 kg) of complete swine feed (Thaler, 2002). Table 6 shows that this substitution would reduce feed costs by $2.10/short ton of finished feed.

Table 6. Economics of adding DDGS to one short ton (907 kg) swine feed.

Ingredients	$/kg	Kg	DDGS cost	Current cost
Add DDGS	0.095	90.7	$8.61	
Add limestone	0.029	1.36	$0.04	
Remove corn	0.093	80.8		$7.51
Remove SBM	0.271	8.63		$2.34
Remove dicalcium phosphate (18.5% P)	0.330	2.72		$0.90
Total cost			$8.65	$10.75
Savings			$2.10	

Generally accepted diet inclusion rates for DDGS

Public and private researchers in the United States have conducted dozens of feed trials over the past 20 years that examine the nutritional implications of feeding various levels of DDGS to beef, dairy, swine, and poultry. Though DDGS inclusion rates vary from one feeding operation to another, Table 7 outlines generally accepted inclusion levels.

---

[11] Based on SBM 89% DM, 49.9% CP; DDGS 92% DM, 30% CP.

Table 7. Generally accepted DDGS diet inclusion levels (Source: various feed trial results; consulting nutritionists).

Species		Maximum % of total ration (DM)
Cattle		
	Lactating dairy cows	20%
	Beef feeders	30-40%
Swine		
	Weaned pigs	25%
	Grow-Finish	20%
	Gestation	50%
	Lactation	20%
Poultry		
	Broilers	15%
	Layers	10%

Corn grain, DDGS and corn gluten supply-demand outlook

Corn

In response to increasing demand, U.S. corn production has dramatically increased in the last eight years. From 1999-2002, annual corn production averaged 242.3 million mt. In contrast, production has averaged 282.8 million mt in the last four years (2003-2006)[12]. In the past three years, corn growers have harvested the largest three crops in U.S. history, including a record 302.6 million mt crop in 2004. Annual corn production is projected to continue steady growth for the next 10 years, driven largely by increased demand from the burgeoning ethanol sector.

Increased production has enabled the U.S. corn industry to satisfy rapidly escalating demand from the ethanol industry, as well as steady demand from the domestic livestock, export, and non-ethanol industrial markets. It is important to note that corn growers have increased production without radically increasing corn acres during this time. This has been possible because of accelerated growth in corn yield per acre. Table 8 shows actual corn supply and demand dynamics for the

---

[12] The year 2006-2007 is estimated based on USDA World Agriculture Supply & Demand Estimates report, Sept. 2006.

Table 8. Corn supply-demand history and forecast, 2005-2006 to 2014-2015. (Source: 99-00 thru 05-06, USDA, 2006; 06-07 thru 14-15, ProExporter Network, 2006).

Crop year	Carry-in 000 mt	Area hvst mil ha	Yield mt/ha	Product 000 mt	Supply 000 mt	Total use 000 mt	Feed use 000 mt	Industrial use 000 mt	Exports 000 mt	Carry-out 000 mt	Farm price US$/mt
05-06	53,699	30.4	9.28	282,261	336,213	285,766	156,854	75,569	53,343	50,447	$79.92
06-07	49,812	29.1	9.71	282,312	332,378	302,582	156,219	91,115	55,248	29,796	$92.51
07-08	30,736	30.9	9.54	294,301	325,291	305,986	139,073	118,650	48,263	19,280	$101.96
08-09	19,280	32.7	10.06	328,745	348,304	329,685	131,452	155,050	43,182	18,619	$102.36
09-10	18,619	32.8	9.96	326,688	345,561	328,009	122,562	164,804	40,642	17,527	$109.05
10-11	17,527	32.9	10.15	334,334	352,115	332,556	123,197	168,716	40,642	19,559	$113.77
11-12	19,559	33.0	10.17	335,426	355,239	332,911	119,387	172,882	40,642	22,328	$121.65
12-13	22,328	33.3	9.56	318,763	341,319	323,284	113,036	169,605	40,642	18,060	$128.34
13-14	18,060	33.0	10.42	343,300	361,615	343,504	116,212	186,650	40,642	18,111	$128.73
14-15	18,111	33.4	11.73	392,173	410,538	375,865	123,197	204,405	48,263	34,673	$129.13

marketing years 1999-2000 through 2006-2007, as well as ProExporter Network's projections through 2014-2015. The table indicates that increased yield and incremental gains in corn acreage are likely to produce a supply of corn that is large enough to satisfy future demand for industrial uses including ethanol without affecting the availability of corn for other uses. It is also relevant to note that an increasing supply of distillers grains will displace more corn in feed rations, freeing up additional supply for other uses. It is estimated that by 2008-2009, more than 39 million mt of corn will be displaced from the U.S. feed market by distillers grains.

Distillers grains

Distillers grains production has doubled in the last five years due to rapid expansion of the ethanol industry. Despite the reported challenges associated with distillers grains consistency and transportation, demand for distillers grains has grown in parallel with available supply. Demand for U.S. distillers grains comes primarily from domestic beef feeders and dairy operators, though usage in swine and poultry markets is increasing. Historically, only a small portion of distillers grains have been exported, though export demand has increased in the past two years. Historical disposition of U.S. distillers grains is illustrated in Figure 5. The majority of distillers grains exports have historically shipped to European Union countries, particularly Ireland and the United Kingdom. However, exports to Mexico, Canada, and Pacific Rim countries have increased considerably in recent years.

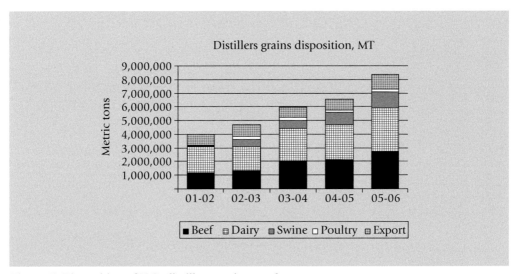

Figure 5. Disposition of U.S. distillers grain supply.

Analysts often express the concern that the U.S. feed market will face an oversupply of distillers grains in the next five to 10 years as dry-grind ethanol production continues to expand, ultimately dragging on plant profitability and return-on-investment for shareholders. Based on the industrial use projections in Table 8, it appears quite feasible that more than 40 million metric tons of distillers grains will be on the market annually by 2014-2015. That amount is more than five times the quantity of distillers grains produced in 2005-2006. Figure 6 shows projections for future distillers grains production.

The theoretical maximum domestic market size for distillers grains is estimated to be approximately 41 million metric tons (Cooper, 2006). This total is based on 100 percent domestic market penetration and the assumption that the annual feed intake of all fed cattle, dairy cows, hogs, layers, and broilers contains the generally accepted maximum level of distillers dried grains. It is highly improbable that distillers grains will ever be integrated into every ration for these animals, highlighting the importance to U.S. ethanol producers of developing a robust export market.

While it is difficult to estimate the theoretical export market potential for DDGS with any level of certainty, it is useful to examine global growth projections for the production of pork, beef, and broiler meat (Tables 10, 11, 12). Expanding production of meat would seem to suggest the potential exists to export more U.S. DDGS to high growth destinations in the future.

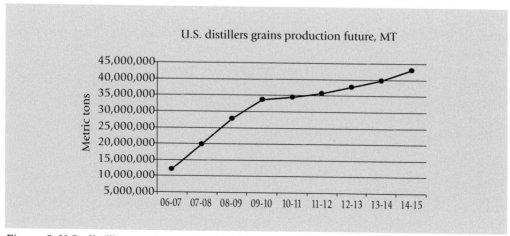

Figure 6. U.S. distillers grains production forecast (ProExporter Network, 2006).

Table 10. Pork production projections, select countries and regions (Food & Agriculture Policy Research Institute, 2005).

Country	2004 (000 mt)	2010 (000 mt)	2014 (000 mt)	% Increase 2004 to 2014
China	47,170	53,733	57,579	22%
EU15	17,850	18,516	19,120	7%
Brazil	2,585	3,229	3,434	33%
Canada	1,900	2,160	2,452	29%
Russia	1,740	2,047	2,139	23%

Table 11. Beef production projections, select countries and regions (Food & Agriculture Policy Research Institute, 2005).

Country	2004 (000 mt)	2010 (000 mt)	2014 (000 mt)	% Increase 2004 to 2014
China	11,178	12,910	13,664	22%
South Korea	6,683	8,523	9,681	45%
Brazil	2,900	3,199	3,479	20%
Australia	2,005	2,246	2,446	22%
Argentina	1,450	1,541	1,673	21%

Table 12. Broiler Production Projections, Select Countries (Food & Agriculture Policy Research Institute, 2005).

Country	2004 (000 mt)	2010 (000 mt)	2014 (000 mt)	% Increase 2004 to 2014
China	9,700	12,065	13,292	37%
Brazil	8,105	9,638	10,608	31%
EU15	6,495	6,984	7,289	12%
Mexico	2,400	2,913	3,263	36%
India	1,650	2,049	2,333	41%

Corn gluten meal and corn gluten feed

Production and use of corn gluten meal (CGM), a co-product of the corn wet mill ethanol process, has remained relatively steady in the past 10 years. Though the rapidly increasing supply of distillers grains may put some downward pressure on CGM demand and prices, supply and demand are projected to remain relatively constant. This is due to well established markets for CGM, reliable supply, and competitive pricing.

In contrast, demand for corn gluten feed (CGF) has decreased steadily in the past 10 years, particularly in the export market. Increased availability of distillers grains is believed to be one reason for the decrease in demand for CGF. Additionally, comparable feedstuffs have become more abundant and economical in previous CGF export markets.

Total export demand in the past nine years for both CGM and CGF is outlined in Figure 7. As discussed previously, dry-grind facilities will be responsible for the overwhelming majority of future growth in the U.S. ethanol industry, so it is unlikely that the available supply of CGM or CGF will increase noticeably.

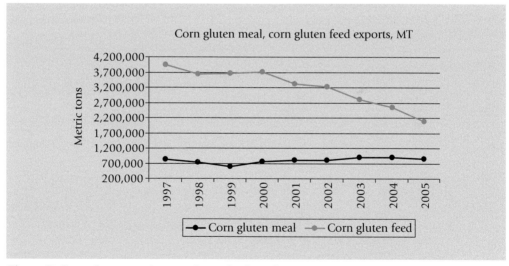

Figure 7. Corn gluten meal, corn gluten feed production and exports (FAS, 2006).

The impact of new ethanol processes on distillers grains

Several new dry-grind ethanol processes are nearing commercialisation or have entered the market in recent years. These new dry-grind processes attempt in many ways to mimic the wet mill process, which essentially divides the corn kernel into its basic nutritional fractions.

Several companies have developed processes that remove the corn germ or corn fiber for the purpose of processing these materials into value-added products. Obviously, removal of these fractions from the kernel at the beginning of the process will have a significant impact on the nutritional composition of the finished distillers grains. The nutrient profiles of several of these new processes are shown in Table 13. While these processes have not yet been widely adopted by the U.S. industry, it is important that the feed industry recognise that considerable volumes of new distillers feed products, with markedly different nutritional profiles, may be on the market in the short term. These new feed products will undoubtedly have novel applications in the feed market and may offer new opportunities for wider inclusion in swine and poultry rations.

Table 13. Sample co-product nutrient profiles resulting from new dry-grind ethanol processes (Source: Company Literature & Technical Presentations).

Nutrient	Renessen	U. of Illinois QG-QF	Broin BFrac (DG HP)	QTI Hydromill (Glutenol)	CPT Dry Frac (Hi-Pro DDGS)
Dry matter, %	n/a	n/a	92.2	90.0	91.0
Crud protein, %	35 - 50	49.31	43.0	45.0	37.0
Fat, %	2.5 - 4.0	4.53	4.3	3.3	5.0
Fiber, %	n/a	6.8	n/a	3.8	5.0
Starch, %	n/a	n/a	n/a	1.5	n/a
Ash, %	n/a	4.13	2.1	4.0	n/a

References

Chase, L. and Linn, J., 1996. Using Distillers Grains in Dairy Cattle Rations. Proc. Alternative Feeds for Dairy and Beef Cattle. p. 13.

Cooper, G., 2006. A Brief and Encouraging Look at Theoretical Distillers Grains Markets. Distillers Grains Quarterly, Spring 06, pgs.13-16.

DOE (U.S. Department of Energy), 2006a. Energy Information Administration, WTI oil prices, tonto.eia.doe.gov/dnav/pet/pet_pri_wco_k_w.htm, accessed September 17, 2006.

DOE (U.S. Department of Energy), 2006b. Energy Information Administration, U.S. regular gasoline prices, all formulations, tonto.eia.doe.gov/dnav/pet/pet_pri_gnd_dcus_nus_w.htm, accessed September 17, 2006.

Food and Agricultural Policy Research Institute, 2005. FAPRI Agricultural Outlook 2006. http://www.fapri.org/outlook2006/

Foreign and Agricultural Service (FAS), 2006. Market and Trade Data. http://www.fas. usda.gov/markettradedata.asp

Klopfenstein, T., 2003. Distillers Grains for Beef Cattle. Nebraska Beef Rep. MP-71-A:32

McClaren, J., 2006. Strathkirn. Presentation to St. Louis Regional Commerce Growth Association.

Nebraska Corn Board, 2006. 2005-2006 Corn Quality Report: 8.

ProExporter Network, 2006. PRX Digest, Sept. 13, 2006.

Reese, D.E. and Lewis, A.J. 1989. Nutrient content of Nebraska corn. Nebraska Cooperative Extension Service EC 89-219, pp. 5-7.

Renewable Fuels Association, 2006. Plant Locations, www.ethanolrfa.org, accessed Sept. 18, 2006.

Shurson, G., and Noll, S., 2005. Feed and Alternative Uses for DDGS. University of Minnesota, Paper presented at Farm Foundation Energy from Agriculture Forum, Dec. 13, 2005. St. Louis, MO. http://www.ddgs.umn.edu/articles-industry/2005-Shurson-%20Energy%20from%20Ag%20Conf.pdf

Thaler, B., 2002. Use of Distillers Dried Grains with Solubles in Swine Diets. College of Agricultural and Biological Sciences, South Dakota State University. ExEx 2035.

Theix, N., 2006. Results of Phase I of Test Method Validation Project, funded jointly by RFA, NCGA, and AFIA. South Dakota State University.

Tilstra, Land O' Lakes/Purina Feed, LLC, 2005. USDA Agriculture Outlook Forum, February 24, 2005, Washington, D.C.

U.S. Department of Agriculture (USDA), 2006. Economic Research Service (ERS), Feed Grains Database, www.ers.usda.gov/Data/Feedgrains/, accessed September 28, 2006.

Yacobucci, B.D. and Womach, J., 2003. Congressional Research Service Report RL30369. Fuel Ethanol: Background and Public Policy Issues. www.ethanol-gec.org/information/ briefing/1.pdf

Developments in the bioethanol market: consequences for the European feed industry

C. Chudaske
Südzucker AG Mannheim/Ochsenfurt, Marktbreiter Str. 74 D-97199 Ochsenfurt Germany; Christine.chudaske@suedzucker.de

Abstract

The EU bioethanol usage will rise considerably. The intra European bioethanol production will need to supply this demand in order to prevent falling short of the goals or becoming dependent on importation. In the EU grain will be used as the main raw material for bioethanol production though sugar or sugar containing substances are also be used. As a consequence Distillers's Dried Grains with Solubles (DDGS) and also vinasses as co-products will reach the market in considerable quantities. Especially the DDGS gains in importance as a protein feedstuff. Worldwide a massive increase in animal production to satisfy the growing need for protein of an ever increasing human population is expected. Concomitant the need for animal feeds will rise too and consequently the requirement particularly for protein rich feedstuffs. Production of bioethanol will change feedstuff markets by supplying DDGS to it. At this moment the degree of self-sufficiency in EU protein feeds is below 30%. Bioethanol will also influence the grain markets in the EU by reducing the oversupply of grains and sugar. This will significantly impact the common agricultural policy as far as intervention and set aside programs are concerned. All together this will lead to changes in the mixture shares of grain in compound feeds and the necessity for a greater diversification of feed materials used in compound feeds. The use and value of co-products as feedstuffs will rise considerably. Good market opportunities for DDGS as a protein rich feedstuff from bioethanol production can be expected.

Keywords: bioethanol, feedstuff, DDGS, grain

Introduction

Recently the developments in the biofuels market are seen in connection with the consequences for agriculture. While crop farmers welcome rising raw materials prices, livestock farmers and the feed industry are less optimistic. Rising the feed stuff market prices of grains and oil seeds might lead to increased feed and tenancy costs. World and also European bioethanol production are rapidly developing. Feedstuffs deriving as co-products from those processes appear in larger quantities on the market. The raw material supply for the European feedstuff industry will be affected significantly by these developments.

History

The advantages of bioethanol as 'the fuel of the future' (so called by Henry Ford) have inspired the pioneers in the automobile and engine construction early on. The use of ethanol played a remarkable role for two essential developments in the modern automobile production. Already in 1860 Nikolaus August Otto used ethanol as fuel for the prototype of his Otto-engine. Almost 50 years later Henry Ford had the 'ethanol' vision. He conceived ethanol to be the real fuel for his 'people car', the legendary T-Model. The series production of this car revolutionised the car manufacturing industry. Ford regarded ethanol as the fuel of the future which simultaneously can give a new growth impetus to farming: *'The fuel of the future is going to come from fruit like that sumach out by the road, or from apples, weeds, sawdust - almost anything'* (New York Times, 1925).

Increasingly the awareness of global warming and the need for a renewable energy supply demands a change in fuel sources. Governmental programs promote increasing the biofuel usage in Europe. The following framework conditions are the main initiatives:

- *Biofuels Directive (2003/30/EC):* Achieve a 2% share of biofuels by 2006 and 5.75% (energy percent) by the end of 2010.
- *Energy Tax Directive (2003/96/EC)*: Member states are allowed to exempt biofuels from energy tax up to 100%.
- *Biomass action plan*: Sets out measures to increase the development of biomass energy from wood, wastes and agricultural crops.
- *EU strategy for biofuels*: Further promotes biofuels in the EU, prepares for the large scale use of biofuels, explores opportunities for developing countries for the production of biofuels factories and biofuels.
- *Green Paper*: EU strategy for sustainable, competitive and secure energy supply.

Bioethanol

Bioethanol is defined as ethanol derived from biological fermentation of biomass e.g. grain or sugar beets. Beside the use for food or industrial purposes about three-quarter of bioethanol is utilised as fuel. The main advantages of bioethanol as a fuel are a reduction in CO_2 emission, lower emission of hydrocarbon, carbon monoxide and particals.

Bioethanol can be used as a direct fuel additive. The EU norm for Otto fuels (DIN EN 228) permits the use of up to 5 Vol%. Also the use of ETBE (Ethyl tertiary buthyl ether; clean burning oxygenate, substitutes Methyl tertiary buthyl ether) as a fuel additive is possible. Flexible Fuel Vehicles are able to use ethanol in petrol with an inclusion rate of 5% up to 85%.

For the production of bioethanol different agricultural products are utilisable. In the EU cereals and sugar beets (Figure 1) are the main raw materials. Sugar beet yields the highest production as well as the highest ethanol production per hectare.

The sugar beet can only be used for the ethanol production after a previous sugar extraction. Momentarily the use of cereals seems to be more competitive than the use of sugar sources in the EU but regional conditions (cultivable area, soil, climate) and the availability of raw materials will play a very important role in the future.

In 2005 approximately 36 million tons of bioethanol were produced worldwide. Compared to this the annual EU production of about 0.4 million tons in 2004 and 0.7 million tons in 2005 is very small. Sharp increases are forecasted however.

Based on the goals for the bioenergy fuels directive of 5.75% (energy equivalent) in 2010 and the fuel consumption prediction for that period, a corresponding need for ethanol can be calculated of up to more than

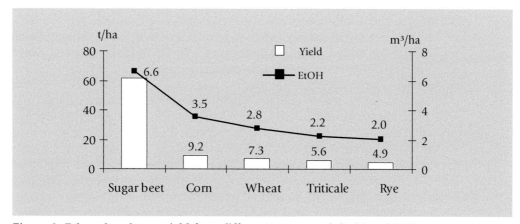

Figure 1. Ethanol and crop yield from different raw materials (Meo, 2003).

8 million tons. Taking into consideration that one ton of ethanol needs about 3 tons of grain, the bioethanol demand will require about 24 million tons of grain per year which is between 8 – 10% of the total EU crop.

Feedstuffs

The projected increase in the EU bioethanol production will also rise the corresponding production of by-products. Distiller's Dried Grains with Solubles (DDGS) is the co-product from bioethanol production from grain. It is used as a protein feedstuff by the compound feed manufacturers as well as for direct feeding on the farm. Vinasses (or Condensed Molasses Solubles) is the co-product from the ethanol production of sugar or molasses and is used in the compound feed production, as substrate for biogas production or as a fertiliser.

The world market for protein feed has grown continuously within the last few years. At present over 200 million tons are traded of which soya is the most important. The forecasts for the coming years predict a market potential of almost 300 million tons in 2010/11. Increases are expected mainly in soya and at other extracted meals or oil cakes. But also the DDGS share will increase further as the market for protein feedstuffs follows the increased total feed consumption. This development is due to the growth in human population and the increasing demand for food. Up to the year 2020 clear increases are predicted primarily of poultry and pig feeds. From this a strong demand especially for protein feedstuffs develops.

In 2010 about 520 million tons of pig feed is needed. This demand will increase by approximately 20% to more than 600 million tons in 2020. To meet this increasing demand of protein sources alternatives to the classic protein feeds like oil cakes and meals are needed. Also increases in production have to be expected but these will, however, lie within the range of about 10%.

The EU feedstuff imports are predominantly protein feedstuffs of which in turn soya is the biggest share (ca. 20 million tons in 2006). The degree of self-sufficiency with protein feedstuffs is very low – only 1 % for soya - and reflects these import shares (Figure 2).

A considerable degree of self-sufficiency is only reached for feedstuffs which numerically don't have great importance such as dried forage. The total degree of self-sufficiency for protein feedstuffs only amounts to about 23%. Changes in the self-sufficiency degree can be expected by the increasing intra-European production of extracted rape seed meal and DDGS. With these changes a reduced dependence on the importation of protein feedstuffs by the amplified production can be reached.

Developments at the feed markets are also caused by the increasing public awareness in industrialised countries for contaminants and

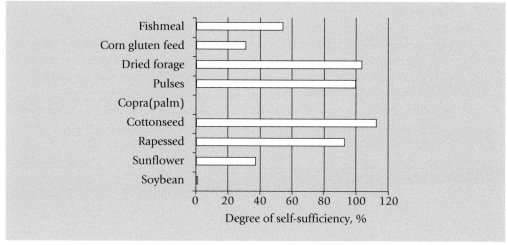

Figure 2. Degree of self-sufficiency with different protein feedstuffs in 2003/2004 (FEFAC, 2006).

consumer health, and the concern for safety, regulation and traceability. This will increase the need for 'safe' feeds from traceable sources and gives market opportunities for standardised 'EU-made feed' from bioethanol production.

Within the last years we have seen an increased usage of grains in compound feeds. The grain share amounted to 52% in 2005/2006 (FEFAC, 2006). The total consumption of grain for feed purposes also rose and almost doubled in the EU since the middle eighties (without considering the EU growth by the addition of new member states). It can be expected that the grain share in the compound feeds will be reduced by the increasing bioenergy production such as bioethanol. The grain markets might change and force a greater diversification of feed materials used in compound feeds. Co-products will reach a high esteem as valuable feedstuffs again.

The compound feed industry's challenge to use these co-products to its nutritional and economical potential will give a pronounced meaning to the phrase value-added industry. They will develop from current 'mixers of grain' back into compound feed manufacturers who can optimally use the co-products in feed mixtures. This optimisation also requires specific use of additives like amino acids or enzymes. Product development trends also have to be expected in this area.

The EU bioethanol production will increase considerably within the next years. Altogether a good marketing potential for DDGS in the EU primarily due to the current low degree of self-sufficiency of protein feedstuffs can be expected. Bioethanol and other biofuels will become a challenge also for the feed industry.

References

FEFAC, 2006. http://www.fefac.org.

Meo, 2003. Bioethanol in Deutschland. FNR e.V., Schriftenreihe NaWaRo 21.

New York Times, 1925. Ford Predicts Fuel from Vegetation. New York Times, Sept. 20, 1925, p. 24.

Speedy, A., 2004. Overview of world feed protein needs and supply. In: Protein Sources for the animal feed industry. Proceedings from FAO Expert Consultation and Workshop, Bangkok, 29 April - 3 May 2002.

By-products from biodiesel production: glycerine and rapeseed products in rations for farm animals

K.-H. Südekum
Institute of Animal Science, University of Bonn, Endenicher Allee 15, 53115 Bonn, Germany

Abstract

This paper briefly reviews utilisation of glycerine as a by-product from biodiesel production and that of other rapeseed products such as rapeseed meal and cake in diets for farm animals with a focus on ruminants. For the benefit of a fail-safe usage of glycerol in diets of all farm animals, methanol should be removed from the glycerine as far as technically possible. Glycerine inclusion levels of 50, 100 and 150 g/kg concentrate dry matter (DM) and storage of concentrates under good (15 °C and 60% relative humidity) or bad (20 °C and 70% relative humidity) environmental conditions for four or eight weeks revealed that physical quality of pellets was not affected by purity of glycerine or by glycerine concentrations of up to 150 g/kg DM. Glycerine at different purities may help to stabilise the hygienic quality of pelleted compound feeds without compromising physical quality of pellets. Data on ruminal turnover of glycerine would suggest that it should replace rapidly fermentable carbohydrates and thus, is not a direct competitor of propylene glycol. Further, it may be speculated that the sweet taste of glycerol may improve intake of diets with inferior palatability (containing, e.g. extensively fermented silages) but this still needs to be investigated. Recent research on rapeseed meal (RSM) has shown that RSM can fully replace soybean meal in dairy cow diets when fed on an approximate isonitrogenous and isocaloric basis, i.e. without considering differences in ruminal degradation and (or) amino acid pattern. Milk and milk component yields were similar for diets containing soybean meal or rapeseed meal. The value of rapeseed cake for all categories of farm animals would benefit from a standardisation of the composition, because varying crude fat and crude protein concentrations makes the feeding value difficult to predict and could also affect storage stability of the cake.

Keywords: glycerine, rapeseed, by-product, ruminant, feeding value

Introduction

The turn towards renewable energy sources in the European Union has increased the production of biofuel from rape seed oil, mainly in the form of rapeseed oil methylester (RME) or biodiesel, leaving glycerine (synonyms: glycerol) as a by-product (see Körbitz et al., 2003; Friedrich, 2004). The approximate proportions of the chemical reaction involved in the production of biodiesel are: (100 l of oil + 10 l of methanol) yield (100 l of biodiesel + 10 l of glycerine; review by Friedrich, 2004).

Starting more than 50 years ago, researchers have shown that glycerine may help to prevent ketoacidosis in the the high-yielding dairy cow by increasing glucose precursors (Forsyth, 1953; Johnson, 1955; Fisher et al., 1971, 1973; Sauer et al., 1973). Recent research on glycerine as a supplement to growing bulls (Pichler and Frickh, 1993) and dairy cows (Khalili et al., 1997; DeFrain et al., 2004; Bodarski et al., 2005) has produced diverging results, thus demanding further labour to elucidate the conditions under which glycerine may be used advantageously as a feedingstuff as opposed to the well-established use as feed additive.

In the first part of this paper, a brief review and summary of data will be presented on glycerine for farm animals with emphasis on ruminants which will encompass the following topics: quality criteria for glycerine, rumen events and energy value. In the second part, again putting an emphasis on ruminants, the feeding value of rapeseed products such as rapeseed meal (solvent-extracted) and rapeseed cake (mechanically extracted; synonym: rapeseed expeller) will be briefly treated, because further increases in the demand for and production of glycerine will also increase the amounts of rapeseed meal and cake.

Glycerine

Quality

Glycerine of varying quality may be found in the marketplace depending on the extent of refinement of the crude glycerine that is obtained during biodiesel production. Table 1 shows data of glycerine of three different purities that were obtained from different stages of the processing of rapeseed oil. It is important to note that the impure quality with elevated methanol concentrations (267 g/kg of dry matter (DM)) was no commodity but an interstage product which was used for experimental purposes only. Almost complete disappearance of methanol occurred when pelleted compound feeds were produced from the glycerines of varying quality at glycerine inclusion levels of 50, 100 and 150 g/kg concentrate DM (Südekum et al., 2007). For the benefit of a fail-safe

Table 1. Chemical composition of glycerine representing different stages of the rapeseed oil methylester production process (Schröder and Südekum, 2002).

Item	Purity of glycerine		
	Low	Medium	High
Water (g/kg)	268	11	25
Dry matter composition (g/kg unless stated)			
Glycerine	633	853	998
Crude fat	7.1	4.4	NA[a]
Phosphorus	10.5	23.6	NA
Potassium	22.0	23.3	NA
Sodium	1.1	0.9	NA
Lead(mg/kg)	3	2	NA
Methanol	267	0.4	NA

[a]NA, not analysed; analyses were omitted because the glycerine content was close to 1000 g/kg and high purity glycerine is listed in the official European Union and United States pharmacopoeias.

usage of glycerol in diets of all farm animals, methanol should be removed from the glycerine as far as technically possible. Lead and other heavy metals were low for all glycerines under investigation.

In Germany, two different glycerines have been listed in the 'Positive List' of feedstuffs as a contribution to feed safety and the specifications are presented in Table 2. Crude glycerine is the quality currently used in farm animal feeding and it is strongly recommended that at least the specifications listed in Table 2 should be given in an instruction leaflet with each batch of crude glycerine.

In the experiment with pelleted compound feeds mentioned above (Südekum *et al.*, 2007) the effects of concentrates, in which glycerine plus soybean meal replaced wheat, on physical, chemical and hygienic pellet quality characteristics were investigated. The three glycerines were as in Table 1, thus reflecting different stages of the RME production process. Inclusion levels in the concentrates of glycerine were 50, 100 and 150 g/kg DM. The quality of the concentrates was assessed under two environmental conditions (good: 15 °C and 60% relative humidity; bad: 20 °C and 70% relative humidity) and two storage durations (four and eight weeks). In summary, chemical composition was only slightly affected by purity and concentration of glycerine or by storage condition

Table 2. Standardised composition of two different glycerine qualities according to the German 'Positive List' (Normenkommission für Einzelfuttermittel im Zentralausschuss der Deutschen Landwirtschaft, 2006).

Item	Glycerine	Glycerine, crude
Glycerine	Minimum 99%	Minimum 80%
Water	0.5 - 1.0%	10 - 15%
Ash	Maximum 0.1%	Maximum 10%
Methanol	0%	Maximum 0.5%
Other	-	NaCl, K, P, S

or duration. Ergosterol, an indicator of fungal biomass, was used to evaluate the hygienic quality of the pellets and data indicated that glycerine of different purities had a preserving effect. Moreover, physical quality of the pellets was not affected by purity or concentrations of glycerine of up to 150 g/kg DM. In conclusion, glycerine of different purities as a by-product from RME production may help to stabilise the hygienic quality of pelleted compound feeds without compromising physical quality of the pellets.

Rumen events

Previous studies on ruminal metabolism of glycerine have indicated that glycerine is extensively fermented in the rumen, propionic acid is the major product of glycerine fermentation (Bergner et al., 1995; Kijora et al., 1998) and that small but significant amounts of glycerine can be absorbed directly from the rumen (K.-H. Südekum, C. Kijora, F. Winden and A. Schröder, unpublished). To further elucidate the fate of glycerine in the rumen, a recent study evaluated effects of glycerine in compound feeds on nutrient turnover in the rumen and digestibilities in the total tract of cattle (Schröder and Südekum, 2002). Four ruminally cannulated steers were used in a 4×4 Latin square design. They were fed on mixed diets (40:60 forage:concentrate, DM basis). The concentrate pellets were isonitrogenous and contained no glycerol or 15% glycerol from glycerol of different purities (63, 85, > 99.5% glycerine in the product). Glycerine intake of the steers was > 1 kg/day.

Total tract digestibilities of organic matter, cell-wall fractions and starch were similar for all dietary treatments. Irrespective of dietary treatment, postprandial pH values in ruminal fluid were always greater than 6.2. The postprandial decline in pH was more pronounced when the diets contained glycerine. Feeding glycerine resulted in a slight shift towards a reduced ratio of acetic acid to propionic acid. Rumen fill (kg) was slightly higher with the diets containing glycerine. The proportion of bailable liquids of total ruminal contents was also higher for the diets containing glycerine of either quality. Obviously, glycerine had an impact on ruminal water turnover. Estimated ruminal *in vivo* fermentation of fibre components was not impaired when glycerine was substituted for starch in the concentrate portion of the diet. Glycerine, when substituted for rapidly fermentable starch sources, e.g. wheat, in ruminant diets should not exert negative effects on ruminal nutrient turnover and digestibilities of organic matter constituents in the total tract. These findings would suggest that glycerine should replace rapidly fermentable carbohydrates and thus, is not a direct competitor of propylene glycol. Further, it may be speculated that the sweet taste of glycerol may improve intake of diets with inferior palatability (containing, e.g. extensively fermented silages) but this still needs to be investigated.

Energy value

Lebzien and Aulrich (1993), based on a digestibility trial on dairy cows, reported that glycerine contains 9.5 MJ/kg net energy for lactation (NEL). Because theoretical considerations would suggest a higher value (11.1 MJ NEL/kg, based on 18.1 MJ gross energy/kg; H. Steingaß, pers. communication), a series of digestibility trials on sheep was conducted to study the effects of purity and concentration of glycerine on energy concentrations of glycerine and on nutrient digestibilities of mixed rations (Südekum and Schröder, 2002). In this study, concentrations of metabolisable energy (ME) and NEL of glycerines as related to purity of glycerine and glycerine content in the DM of mixed diets were derived from digestibility trials with wethers. Glycerine of varying purity (63, 85, > 99.5% glycerine in the product) was mixed with forage (wilted grass silage) and a high-starch concentrate. The mixed diets were formulated to contain (DM basis) 40% forage, 50% concentrate and 10% of pure glycerine irrespective of the purity of the glycerine-containing product. A 100% forage diet and a glycerine-free forage:concentrate (40:60) diet served as controls. In addition, the glycerol of the highest purity (> 99.5% glycerine) was included in diets containing 40% forage and 5, 10, 15 or 20% glycerine. The high-starch concentrate or a low-starch concentrate made up the balance of these diets (55 to 40% of the dry matter). The NEL concentrations were higher when glycerine was fed with the low-starch concentrate than when fed with the high-starch concentrate (Table 3).

The estimated ME and NEL concentrations for glycerol derived from diets containing the low-starch concentrate and 10, 15 or 20% glycerol in the dry matter were similar. When fed with the low-starch concentrate, either no or positive effects of glycerol on nutrient digestibilities (organic

Table 3. Concentrations of metabolisable energy (ME) and net energy for lactation (NEL) of pure glycerine estimated from different proportions of glycerine in the dry matter of rations containing high-starch or low-starch concentrates (Südekum and Schröder, 2002).

Glycerine (% of diet dry matter)	Concentrate type			
	High-starch		Low-starch	
	ME	NEL	ME	NEL
5	14.4	9.6	11.1	6.9
10	13.1	8.4	14.6	9.7
15	12.6	8.0	14.9	9.9
20	13.2	8.5	14.4	9.5
SEM[a]	0.9	0.7	0.9	0.7

[a]SEM, standard error of the means.

matter, starch, cell-wall components) were observed. When derived from diets containing the high-starch concentrate as opposed to the low-starch concentrate, glycerol of different purities and at different dietary inclusion levels (10, 15 or 20 %) had markedly lower NEL concentrations (8.0 to 8.5 MJ/kg). Similarly, digestibilities of cell-wall components were depressed yet without reducing organic matter digestibilities.

In conclusion, glycerines of different purities can be included in mixed diets up to 10% of the DM of the diet without negatively influencing nutrient digestibilities. When fed with a low-starch concentrate, pure glycerine at dietary inclusion levels up to 20% had no or positive effects on nutrient digestibilities. When included in diets containing high-starch concentrates, however, glycerine reduced cell-wall digestibilities but had no obvious effect on organic matter digestibilities.

In summary, the net energy value for ruminants of glycerine was similar to previously published values for dairy cows (9.5 MJ NEL/kg; Lebzien and Aulrich 1993) but considerably lower than derived from theoretical considerations (11.1 MJ NEL/kg, based on 18.1 MJ GE/kg; H. Steingaß, pers. communication). The difference between the two values may be due to ruminal fermentation, post-absorptive metabolism and(or) limitations of standard digestibility trials to estimate ME and NEL values. It is noteworthy that poultry and pig data for ME values (17.5 - 17.6 MJ/kg;

Bartelt and Schneider, 2002) better match theoretical considerations for ruminants (16.3 MJ/kg; H. Steingaß, pers. communication) than the ruminant data itself.

(Other) rapeseed products: rapeseed meal and rapeseed cake

The protein values of soybean (SBM) and rapeseed (RSM) meals published in feeding value tables in the past differed significantly. The proportion of the crude protein (CP) that is ruminally undegraded (RUP) was 35% for SBM, significantly higher than the 25% stated for RSM (Universität Hohenheim – Dokumentationsstelle, 1997). More recent experiments indicated that above all the large differences between the tabulated ruminal degradability values of the two meals in favour of SBM no longer reflect today's qualities. Data on in situ ruminal degradation rate are shown in Table 4 and indicates that the CP of SBM is degraded even more rapidly than that of RSM. This data is based on a cross-sectional study conducted in Germany covering all oil mills processing rapeseed and soybeans and in addition encompassing some imported SBM commodities. Table 5 provides comparative values for RUP and the utilisable crude protein (uCP) at the duodenum, derived from a comparison of in situ and several *in vitro* methods on the 17 SBM and RSM of the crosss-sectional study. An overall view of all our own findings and those published by other working groups results in the following conclusions:
1. RSM contains more and SBM less RUP than previously stated.
2. The smaller difference (by comparison with the values in the tables) in uCP content between SBM and RSM is attributable to the changes in the proportion of RUP.

Table 4. Rates (%/h) of in situ ruminal degradation of crude protein of rapeseed (RSM) and soybean (SBM) meals[a] (Südekum et al., 2003).

	RSM	SBM
Mean	12	16
RSM vs. SBM	P < 0.0001	
Maximum	19	19
Minimum	3	10

[a]10 RSM (German oil mills) and 7 SBM (German (n = 4), Dutch, Argentine and Brazilian oil mills).

Table 5. Protein value of contemporary qualities of rapeseed (RSM) and soybean (SBM) meals as compared with feeding table values (Südekum et al., 2003).

	RSM	SBM
Mean RUP[a], % of crude protein	30	30
DLG Table (Universität Hohenheim – Dokumentationsstelle, 1997)	25	35
Mean uCP[b], g/kg dry matter	231	288
DLG Table (Universität Hohenheim – Dokumentationsstelle, 1997)	219	298 – 308

[a]RUP, ruminally undegraded crude protein.
[b]uCP, utilisable crude protein at the duodenum (sum of microbial and ruminally undegraded crude protein).

Since the reference values determined in vivo on duodenally cannulated dairy cows from selected meals largely but not completely confirm the very clear findings from other methods, it was suggested that for the time being for RSM and SBM a mean RUP proportion of CP of 30% should be stated and the uCP values should be correspondingly adjusted. The DLG, the German body responsible for documenting feeding values, accepted this proposal and published it on the internet under http://www.dlg.org/de/landwirtschaft/futtermittelnet/index.html.

At the same time when studies evaluated the fate of CP of SBM and RSM during ruminal degradation, other researchers in a number of applied research stations throughout Germany tested the hypothesis that RSM can fully replace SBM in dairy cow diets when fed on an approximate isonitrogenous and isocaloric basis, i.e. without considering differences in ruminal degradation and (or) amino acid pattern. Data summarised in Table 6 indicate that in the four trials, milk yield and milk component concentrations were similar for diets containing SBM or RSM and thus, the hypothesis can still be sustained.

In conclusion, it should be pointed out that the overall quality of RSM and rapeseed cake depends also on the concentrations of glucosinolates and in the case of rapeseed cake, the content and quality of the lipid portion. Table 7 would indicate that the average glucosinolate concentrations of RSM are low and that of rapeseed cake are considerably higher and that both feedstuff types show a considerable variation for this item. Also, crude fat in rapeseed cake varies largely, making ration programming a difficult task without having analytical data of specific batches at hand. Increasing crude fat contents lowers crude protein concentration and vice versa. Grouping of rapeseed cakes according to crude fat

Table 6. Comparative evaluation of rapeseed (RSM) and soybean (SBM) meals in diets for high-producing dairy cows - summary of German trials (Spiekers and Südekum, 2004).

Location, duration of trials and diets	Protein supplement kg/(day x cow)	Milk kg/day	Fat %	Protein %
LWZ Haus Riswick: lactation weeks 5 - 35				
Basal diet	SBM 2.3 kg	31.1	3.9	3.1
1/3 MS[a] + 2/3 GS[b]	RSM 3.1 kg	31.3	3.9	3.2
LWZ Haus Riswick: lactation weeks 2 - 44				
TMR with	SBM 1.6 kg	25.2	4.2	3.4
50% MS + 25% GS	RSM 2.2 kg	25.8	4.1	3.4
LVA Iden: until lactation week 17				
TMR with 40% (MS +	SBM 4.0 kg	40.0	3.8	3.3
ECS[c]) + 25% GS	RSM 4.3 kg	40.5	3.9	3.3
LVA Köllitsch: 17 weeks				
Basal diet	SBM 1.6 kg	31.2	3.9	3.4
50% MS + 50% GS	RSM 2.0 kg	32.7	4.0	3.4

[a]MS, maize silage.
[b]GS, grass silage.
[c]ECS, ear corn silage.

concentration appears necessary, e.g., 7 - 10%, 12 - 15%, other (greater values are often found in small-scale enterprises). Further, storage stability should be considered because the fat is in a non-protected form after mechanical extraction of the seed. It has also been reported by farmers and consultants that physical characteristics resulting from plaque forming during oil extraction make rapeseed cake difficult to handle and that a homogenous distribution in totally mixed rations (TMR) or silage mixtures is difficult to achieve.

It becomes also clear from this data that a more widespread use of RSM and rapeseed cake in diets for pigs and poultry requires further reduction of glucosinolate levels, particularly for the cake, and lesser variations would also be helpful in this regard. The crude fat content of the cakes should also be more standardised to make use of the commodities easier and more reliable.

Table 7. Quality of rapeseed cake and rapeseed meal – survey data from Germany (miscellaneous sources; Weiß, 2007).

Type and source of feed	n =	GSL[a] (mmol/kg DM[b]) (Min - Max)	Crude fat (% DM) (Min - Max)
Rapeseed meal (solvent-extracted)			
10 German oil mills	637	8.3 (1 - 20)	
Monitoring 2006 (UFOP)	68	8.2 (4.4 - 11.2)	
Rapeseed cake (mechanically extracted)			
6 plants	85	22.1 (15 - 29)	12.6 (9 - 17)
31 local plants	94	15.9 (7 - 28)	15.1 (9 - 28)
22 local plants	22	13.5 (5 - 22.4)	16.9 (12.9 - 24.3)

[a]GSL, glucosinolates.
[b]DM, dry matter.

Table 8 summarises current German recommendations for rapeseed products in diets of cattle and pigs to the best of one's knowledge. Pigs would particularly benefit from progress in further reduction of glucosinolate levels whereas in cattle, a safer quality assessment of the rapeseed cake is needed from which also pigs would benefit.

Table 8. Practical recommendations for daily amounts or dietary concentrations (as fed basis for dry diets) of rapeseed products for cattle and pigs (Weiß, 2007).

Animal category	Rapeseed meal, solvent-extracted	Rapeseed cake, mechanically extracted
Dairy cow	Maximum 4 kg	1.5 - 2.0 kg
Beef cattle	Maximum 1.2 kg	1 kg
Fattening pigs	Maximum 10%	7 - 10%
Sows	5 - 10%	5 - 10%
Piglets	-	5 - 10%

Conclusions

Glycerine is a versatile feedingstuff in particular for ruminants but is different from propylene glycol. Mature cattle can consume (and like!) considerable quantities of glycerine (1 kg/d). The net energy (NEL) concentration is approximately 9.5 MJ NEL/kg of glycerine. Conflicting results from trials on dairy cows indicate that more research is necessary to define conditions that allow glycerine to be used advantageously.

Other rapeseed products for ruminants, such as rapeseed meal, compare well with soybean meal for dairy cows if fed on an isonitrogenous basis. Rapeseed cake needs further consideration and more reliable data because variations in the processing conditions result in varying chemical composition, particularly regarding the crude fat content and this currently hampers the prediction of its feeding value for all categories of farm animals.

References

Bartelt, J. and Schneider, D., 2002. Untersuchungen zum energetischen Futterwert von Glycerol in der Fütterung von Geflügel und Schweinen. UFOP-Schriften 17, 15-36.

Bergner, H., Kijora, C., Ceresnakova, Z. and Szakács, J., 1995. In vitro Untersuchungen zum Glycerinumsatz durch Pansenmikroorganismen. Arch. Anim. Nutr. 48, 245-256.

Bodarski, R., Wertelecki, T. Bommer, F. and Gosiewski, S., 2005. The changes of metabolic status and lactation performance in dairy cows under feeding TMR with glycerin (glycerol) supplement at periparturient period. Electronic J. Polish Agr. Univ. 8 (4), 9 pp. Accessible online at: http://www.ejpau.media.pl/volume8/issue4/art-22.html.

DeFrain, J.M., Hippen, A.R., Kalscheur K.F. and Jardon, P.W., 2004. Feeding glycerol to transition dairy cows: Effects on blood metabolites and lactation performance. J. Dairy Sci. 87, 4195-4206.

Fisher, L.J., Erfle, J.D., Lodge G.A. and Sauer, F.D., 1973. Effects of propylene glycol or glycerol supplementation of the diet of dairy cows on feed intake, milk yield and composition, and incidence of ketosis. Can. J. Anim. Sci. 53, 289-296.

Fisher, L.J., Erfle, J.D. and Sauer, F.D., 1971. Preliminary evaluation of the addition of glucogenic materials to the rations of lactating cows. Can. J. Anim. Sci. 51, 721-727.

Forsyth, H., 1953. Glycerol in the treatment of (1) bovine acetonemia, (2) pregnancy toxaemia in ewes. Vet. Rec. 65, 198.

Friedrich, S., 2004. A world wide review of the commercial production of biodiesel – a technological, economic and ecological investigation based on case studies. Schriftenreihe Umweltschutz und Ressourcenökonomie, Vol. 41. Institut für Technologie und nachhaltiges Produktmanagement, Wirtschaftsuniversität Wien, Austria, 150 pp.

Johnson, R.B., 1955. The treatment of ketosis with glycerol and propylene glycol. Corn. Vet. 44, 6-21.

Khalili, H., Varvikko, T., Toivonen, V., Hissa, K. and Suvitie, M., 1997. The effects of added glycerol or unprotected free fatty acids or a combination of the two on silage intake, milk production, rumen fermentation and diet digestibility in cows given grass silage based diets. Agric. Food Sci. Finland 6, 349-362.

Kijora, C., Bergner, H., Götz, K.-P., Bartelt, J., Szakács, J. and Sommer, A., 1998. Research note: Investigation on the metabolism of glycerol in the rumen of bulls. Arch. Anim. Nutr. 51, 341-348.

Körbitz, W., Friedrich, S., Waginger, E. and Wörgetter, M., 2003. Worldwide review on biodiesel production. Austrian Biofuels Institute, Wieselburg, Austria, 121 pp.

Lebzien, P. and Aulrich, K., 1993. Zum Einfluss von Glycerin auf die Rohnährstoffverdaulichkeit und einige Pansenparameter bei Milchkühen. VDLUFA-Schriftenreihe 37, 361-364.

Normenkommission für Einzelfuttermittel im Zentralausschuss der Deutschen Landwirtschaft, 2006. Positivliste für Einzelfuttermittel (Futtermittel-Ausgangserzeugnisse), 5th ed. Zentralausschuss der Deutschen Landwirtschaft, Berlin, Germany. Available online at: http://www.futtermittel.net/pdf/positivliste_5.pdf.

Pichler, W.A. and Frickh, J.J., 1993. Der Einsatz von Glycerin aus der Rapsölmethylestererzeugung in der Jungstiermast. Förderungsdienst 41 (4), 25-28.

Sauer, F.D., Erfle, J.D. and Fisher, L.J., 1973. Propylene glycol and glycerol as a feed additive for lactating dairy cows: An evaluation of blood metabolite parameters. Can. J. Anim. Sci. 53, 265-271.

Schröder, A. and Südekum, K.-H., 2002. Effekte von Glycerin unterschiedlicher Reinheit auf die Pansenfermentation und Nährstoffverdaulichkeiten bei Rindern. UFOP-Schriften 17, 51-67.

Spiekers, H. and Südekum, K.-H., 2004. Einsatz von 00-Rapsextraktionsschrot beim Wiederkäuer. UFOP-Praxisinformation. UFOP, Berlin, 17 pp.

Südekum, K.-H., Nibbe, D., Lebzien, P., Steingaß, H. and Spiekers, H., 2003. Comparative evaluation of the protein values of soybean and rapeseed meals by in vivo, in situ and laboratory methods. In: Proc. 11th Int. Rapeseed Congr., Towards enhanced value of cruciferous oilseed crops by optimal production and use of the high quality seed components. Copenhagen, Denmark, 1241-1243.

Südekum, K.-H. and Schröder, A., 2002. Einfluß der Reinheit und Konzentration von Glycerin auf die Energiegehalte von Glycerin und die Nährstoffverdaulichkeiten gemischter Rationen für Wiederkäuer. UFOP-Schriften 17, 37-50.

Südekum, K.-H., Schröder, A., Fiebelkorn, S., Schwer, R. and Thalmann, A. 2007. Quality characteristics of pelleted compound feeds under varying storage conditions as influenced by purity and concentration of glycerol from biodiesel production. J. Anim. Feed Sci. (in press).

Universität Hohenheim – Dokumentationsstelle (editor), 1997. DLG-Futterwerttabellen Wiederkäuer. 7th edition. DLG, Frankfurt/Main, Germany.

Weiß, J., 2007. Futtermittel aus der Rapsverarbeitung. In: Christen, O. and Friedt, W. (editors), Winterraps - Das Handbuch für Profis. DLG, Frankfurt, Germany, pp. 277-289.

Applications of rapeseed meal or –expeller and glycerine in diets for non ruminants

J. Doppenberg and P.J. van der Aar
Schothorst Feed Research B.V., P.O. Box 533, 8200 AM Lelystad, the Netherlands; jdoppenberg@schothorst.nl

Abstract

Biodiesel production is expected to increase significantly in the near future. Two by-products become available for the feed industry rapeseed meal or -expeller and glycerine. The glucosinolate (GSL) content of rapeseed products is the main anti nutritional factor to be considered for formulating swine feeds. With a GSL content of about 10 µmol GSL/gram 'OO' rapeseed meal can be included in growing/finishing swine feeds up to 20-25%. The GSL-level of rapeseed expellers is higher. The digestibility of rapeseed products depends on the target animal and the degree of feed processing. Glycerine is a liquid by-product that as an energy source up to 5% in swine diets and 10% in poultry diets efficiently (and with current prices economically) can be used. At higher inclusion rates excretion via the kidney increases, leading to energy losses and wet litter. In finishing pigs glycerine improves the water binding capacity of meat and meat products.

Keywords: glucosinolate, rapeseed meal, rapeseed expeller, glycerine, energy value

The nutritional value of by-products from biodiesel production: rapeseed meal or-expeller

Introduction

Increased bioenergy production will increase the amount of plant protein that becomes available for the feed industry. The main by-products from both bioethanol (DDGS from corn, barley or wheat) and biodiesel production (sunflower meal, rapeseed meal or –expeller) are both high in protein (30% protein on a dry matter base). Therefore the protein prices are likely to decline. Increased bioenergy production in the EU also means that the need to import plant protein for animal production in the form of, e.g. soybeans will decline. Both sunflower- and rapeseed can be used for biodiesel production, however the climate in he Northern Hemisphere is more suitable for rapeseed production. Rapeseed meal and –expeller are nutritionally more advantageous for non ruminants because of the lower fiber and higher energy content than sunflower meal.

Rapeseed meal and –expeller

For a few key nutrients the chemical composition of rapeseed, rapeseed meal and –expeller are outlined in Table 1.

Rapeseed (*Brassica napus* and *campestrus*) contains 42-44% fat (CVB and INRA, 2004). Rapeseed meal is a by-product from the more efficient steam/solvent extraction process leaving only 2-4% residual fat. Rapeseed expeller is a by-product from the cold pressing process. The fat content varies from mill to mill (and from batch to batch); it can be as low as 7% and as high as 20%. The variable fat content of rapeseed expeller demands the need for constantly monitoring the product and readjusting the feed formulations accordingly.

The fatty acid pattern of rapeseed oil is characterised by a high oleic acid (C18:1) content. It contains 60% C18:1 of fatty acids. Only olive oil has a higher oleic acid content with 72%. Consequently the linoleic acid (C18:2) is low with 20% of fatty acids, compared to corn and soy oil. It contains a fair (10% of fatty acids) amount of linolenic acid (C18:3), resulting in a low omega 6:omega 3 fatty acid ratio.

Essentially seeds rich in oil are low in starch and sugars. After oil extraction a protein and fiber rich by-product is left. The fiber fraction is relative high in the ADF fraction and low in NDF or water soluble fiber. This decreases the digestibility of the fiber fraction for non ruminants but on the other hand also decreases the probability of causing wet litter with poultry and diarrhea with pigs.

The protein quality in terms of essential amino acid content and (ileal) digestibility of rapeseed products is less than that of soy protein. The ileal digestible amino acid content of rape seed protein is better than that of sunflower protein. Ideally feeds therefore need to be optimalised on an (ileal) digestible amino acid base when using different protein sources.

Table 1. Chemical composition rapeseed products.

In product	Rapeseed	Rapeseed meal	Rapeseed expeller
Crude protein %	20-22	32-35	31-34
Crude fat %	42-44	2.5-3.5	7-13
Crude fiber %	7-10	11-13	10-15
Lys/ 16 g N	5.4-5.8	5.4-5.8	5.4-5.8
Dig. lysine %	70-78	72-80	72-79

Anti nutritional factors in rapeseed, rapeseed meal and –expeller

Native varieties of mustard seed (*Brassica juncea*) can contain very high levels (up to 50% of fatty acids) of erucic acid (22:1). Erucic acid is very toxic to humans and can be carried over in animal products like milk. Single zero ('O') varieties of rapeseed have been developed as early as 1968, which contain very low levels of erucic acid. The erucic acid content of modern varieties is about 0.2% of fatty acids. Since rapeseed meal and to a lesser extend rapeseed expeller are low in fat, erucic acid is of little concern in the 'O' varieties.

With layers the sinapin(e) content is of particular concern for brown layers. The average sinapine content of rapeseed meal is about 1%. Brown layers lack the TMA-oxidase enzyme so that trimethylamine (TMA) formed from sinapine is not metabolised and can be transferred in eggs causing an off-flavour (fishy smell). With all poultry the relatively high fiber and low energy content of rapeseed meal limits its usage. Technological treatments like pelletising and/or expending the feed improves the rate of digestion mainly through a reduction of particle size and agglomerising starch.

The main nutritional concern of rapeseed products for swine feeds is the glucosinolate (GSL) content. The group of glucosinolates (or thioglucosides) consists of more than 100 different compounds. They have in common that they all contain a sulfur and a glucose group. Depending on the rest group they are further defined as, e.g. sinigrin, progoitrin, sinalbin, glucobrasscin, etc.

Myrosinase, a natural occurring enzyme in rapeseed, becomes active during handling and processing (grinding) of rapeseed. The first step is to split of the sulfur and glucose group. Further metabolisation of the rest group leads to toxic products like isothiocyanates (ITC), vinylthiooxazolidines (VTO) and nitrils. These components decrease the uptake of iodine by the thyroid gland, limiting the production of thyroid hormones (hypothyroidism). Thyroid hormones are necessary for normal growth. The thyroid compensates for it by increased cell division leading to hypertrophy also known as goiter. ITC has a pungent taste reducing the feed intake. The liver can detoxify these components but increasing GSL-concentrations lead to liver enlargement. An overload of toxins leads ultimately to liver damage ('fatty livers') and an inability to perform normal liver functions necessary for metabolism. High GSL-levels in grower feeds result in a growth reduction and in sow feeds in an increase in embryonic losses during early gestation.

'OO' rapeseed meal

So called 'OO' (double zero) varieties of rapeseed were developed, which contain 5-10 x lower GSL-levels than 'O' (single zero) rapeseed or earlier varieties of mustard seed. In Canada this product is called canola. Figure 1 shows that the GSL-content of canola meal continues to decline (DeClerq, 2005). However it can vary between 4-31 µmol/g.

The Canadian Grain Commission reports an average GSL content of 11 µmol/g for canola (DeClerq, 2005). The Commission reports an average of 16 µmol/g for canola and mentions that traditional rapeseed meal contained between 120 and 150 µmol/g of total glucosinolates (Hickling, D. 2001). In the EU almost all rapeseed is of the 'OO' (double zero) variety. In Asia native varieties of rapeseed with high levels of GSL are likely to be found, increasing the need to monitoring GSL-levels of received products.

In the EU glucosinolates are often expressed as the metabolites ITC (M=115) and VTO (M=129) in mg/g. VTO is considered to be more damaging, since the iodine binding by VTO can not be off set by additional iodine supplementation. Moreover the very toxic nitrils are produced formed VTO. The average ITC content of 'O' rapeseed meal is 2-3 g/kg and the VTO content is 4-5 g/kg, these levels are 0.3-0.4 g/kg ITC and 0.6-1.2 g/kg VTO in 'OO' varieties of rapeseed meal.

German research suggest that there is a significant difference in the GSL-content of different rapeseed varieties and within the same variety. Values between 7.0 and 28.6 µmol/g are reported, the yearly average varies from 12.7 to 14.5 µmol/g (Schuman, 2005). Südekum (2006) further shows that there is a significant difference in the GSL-content of rapeseed meal which is heat treated and rapeseed expeller which is cold

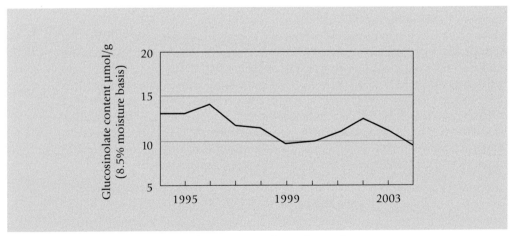

Figure 1. Average yearly glucosinolate content of canola meal.

pressed. While the average GSL content of rapeseed meal is 8.3 µmol/g, that of rapeseed expellers (13-17% fat) is twice as high. Consequently, processing of feedstuffs and feed has an influence on the GSL content and ultimate usability of rapeseed products in rations.

At Schothorst Feed Research a trail was conducted with growing/ finishing pigs fed *ad libitum* experimental feeds with an increasing amount of glucosinolates (Borggreve and Veldman, 1989). Both 'O' and 'OO' varieties of rapeseed meal were used at increasing inclusion rates in order to raise the GSL of the feeds. Rapeseed meal replaced soybean meal and sunflower meal. An *ad libitum* feeding regime was chosen to study the effect of the GSL content of the feed on feed intake. Pigs were fed experimental diets from 25-107 kg B.W. and housed individually. The results are given in Table 2.

Glucosinolate concentrations of 2.2 µmol GSL/gram of feed or higher and/or addition levels of 4% or more of 'O' rapeseed meal significantly reduced the feed intake and average daily gain. The average daily gain was reduced in a linear fashion indicating that the increased toxin load negatively influenced the metabolism of nutrients. The thyroid weight at slaughter also responded to increasing GSL concentrations in a significant and linear fashion. Liver weights were significant increased with GSL concentrations of 4.4 µmol GSL/gram and higher compared to the control.

Histological work showed an increase in hormone (T3 and T4) producing cells in the follicles of the thyroid. No liver damage or occurrence of fatty livers was observed at any GSL levels. Also no circulating liver enzymes like SDH (sorbitol dehydrogenas), LDH (lactate dehydrogenase) or GOT (alanine transferase) were found in the plasma.

Table 2. The effect of GSL concentrations of feed intake and growth of swine.

% rapeseed meal	Ctrl	'OO' 12%	'O' 4%	'O' 8%	'O' 12%	LSD 95%
GSL µmol/g	0	1.4	2.2	4.4	6.6	
Feed kg/d	3.06	3.00	2.87	2.93	2.82	0.13
ADG g/d	953	947	916	908	885	58
FCR kg/kg	3.22	3.20	3.17	3.23	3.22	0.14
Thyroid gram	9.6	13.1	16.6	23.1	28.1	4.4
Liver gram	1,862	1,933	1,877	2,244	2,237	154

In conclusion:
1. GSL levels of 2.2 µmol GSL/gram or more cause hyperthrophy of the thyroid.
2. GSL levels of 2.2 µmol GSL/gram or more reduce feed intake.
3. GSL levels of 4.4 µmol GSL/gram or more cause hyperthrophy of the liver.
4. GSL levels of 6.6 µmol GSL/gram or more reduce average daily gain.

In follow up experiments (Borggreve and Veldman, 1989; Fledderus et al., 2000) pigs were fed restricted according to a practical feeding schedule. No effects on feed intake with increasing GSL concentrations up to 4.7 µmol/g could be observed. The liver weight was again significantly increased at this level and the growth tended to be lower. These results are in agreement with a recent excellent review article in which the authors concluded that the total glucosinolate content of swine feeds should be kept below 2 µmol GSL/gram and supplemented with 1 mg/kg Iodine. Of all livestock species pigs are more sensitive to GSL than poultry, rabbits or fish and ruminants are the least sensitive (Tripathi and Mishra, 2007).

It can therefore be concluded that 'OO' rapeseed meal, that contains about 10 µmol GSL/gram can be included in growing/finishing swine feeds up to 20-25%.

The energy value of rapeseed products

The relative value of rapeseed expellers depends on the age of the animal. When an animal matures and the digestive capacity develops, in particular the large intestine, the ability to ferment complex carbohydrates increases. Fermentation can significantly add to the energy value of high fiber products like, e.g. the by-products from bioenergy production (DDGS and rapeseed meal/expeller) for growing/finishing pigs and sows. This results in a different ranking among feedstuffs and use in formulations for each animal category, based on energy value and price.

Schothorst Feed Research has developed different Net Energy tables for piglets, fattening swine and gestating sows. In Table 3 the relative energy value of rapeseed expeller is given relative to the energy content of corn with different energy systems and sources.

In a Metabolisable (ME) or Digestible Energy (DE) system the value of protein rich feedstuffs is overestimated. The Dutch CVB and French INRA tables have lower NE-values than Schothorst Feed Research has determined, because they do not take the effect of processing (pelleting) of feed in consideration. Heat processing agglomerates starch increasing viscosity in the gut and pelleting decreases particle size aiding the digestion of feeds. Note that corn itself (a high starch feedstuff) has a

Table 3. Energy value in relation to the age of animals and the energy system.

Energy system	ME	NE			
Animal category	Pigs	Pigs	Piglets	Gr/fin	Gest. sow
Reference	SFR	CVB	SFR	SFR	SFR
Corn	100	100	100	100	100
Rapeseed exp.	92.1	72.4	76.0	80.3	83.7

different energy value for each animal category. The differences between corn and rapeseed expeller are most severe with younger animals.

In order to obtain consistent results with the by-products from the bioenergy production in feed formulations the nutritional value needs to be estimated based on digestibility parameters (energy and amino acids) measured with each target animal category.

The nutritional value of by-products from biodiesel production: glycerine or glycerol

Introduction

The names glycerin, glycerine or glycerol are used synonymous. Glycerine is a liquid high energy product that becomes available by the hydrolyses of fatty acids from (animal) fat or pure plant oils (PPO). Plant oils used for biodiesel production are mostly derived from sunflower- or rapeseed. The use of fatty acids rather than PPO for biodiesel production prevents the need for modifications to the engine. Fats and oils are chemically defined as triglycerides which contain about 10% glycerine and 90% fatty acids. Increased biodiesel production means that an enormous surplus of glycerine will become available, significantly lowering the glycerine prices. Estimates for 2010 amount to one million metric ton of glycerine (Rick, 2006).

Increased bioenergy production will increase the cost of the energy components of animal feeds. First of all less arable land will be available for crop production because of competition with biomass production. Secondly starch (grains) and sugar rich (sugar beets and molasses) feedstuffs will increasingly be used for bioethanol production. The main by-products from both bioethanol (DDGS from corn, barley or wheat) and biodiesel production (rapeseed meal or –expeller) are both high in protein (30% protein on a dry matter base) but relative low in energy.

Therefore glycerine becomes an attractive energy rich feedstuff for the feed manufacturing industry, within the food feed or fuel competition.

Production process and residues

The pharmaceutical grade of glycerine contains 99.9% glycerol. Almost all the research quoted in this paper has been done with that product. In certain countries (Germany) only glycerine derived from pure plant oils are registered as approved feedstuffs. Potentially the risk of undesired residues (e.g. dioxins) can be higher with glycerine derived from animal or waste fats.

The feed grade glycerine contains a minimum of 80% glycerine, the remainder is mostly water and minerals. For the hydrolyses of triglycerides (oils) lye (NaOH or KOH) are used as catalysts and methanol for methylation. Methanol is retained by distillation, however residues (max. 0.5%) are found in the feed grade product. The lye is neutralised with hydrochloric acid (HCL), resulting in a high salt (NaCL or KCL)-content. The sodium content is minimal 2% and the ash content maximal 7.5% (Rick, 2006).

The high salt content will maximise the usability of the feed grade glycerine to 5% in swine and layer feeds and 10% in broilers feeds and dairy concentrates, even by replacing added salt.

Methanol toxicity

Methanol has a boiling point of 64,5 °C, consequently residual methanol will be evaporated during processing (pelleting) of feeds. Südekum and Schröder (2006) found no negative effects on rumen fermentation or energy retention by using in a pelleted concentrate a glycerine product that contained 26.7% methanol. Methanol vapours can however pose a hazard in the feed mill. Methanol is likely to be fermented in the rumen instead of being absorbed.

In practise the methanol content of the feed grade glycerine will be ± 0.01%. It can be calculated that with normal feed intake levels and 3% glycerine addition plasma methanol levels will remain below 0.2 mg/dl. Kostic and Dart (2003) reports 20 mg methanol/dl plasma to be the threshold level for hospitalisation of humans. Extrapolating these values the safety margin will be 100x.

Prabhakaran et al. (1993), moreover, reports that the half life of methanol in humans is only 2 to 24 hours and that is eliminated via the lungs (10-20%) and the kidneys (3%). Not methanol itself but the metabolites formed during oxidation are toxic. Formic acid accumulation can lead to blood acidosis and unconsciousness. The initial activation step is alcohol dehydrogenase (ADH) dependant. This enzyme favours ethanol. Therefore, remarkably, ethanol can be used effectively in the treatment of methanol poisoning to prevent oxidation of methanol (Prabhakaran *et al.*, 1993).

The eventual risk of methanol toxicity by feeding methanol containing feed grade glycerine to animals or humans consuming animal products will be minimal and limited to meal fed feeds to non ruminants.

Glycerine applications

Glycerine is hygroscopic. The water binding capacity has been used as a preservative (mould inhibitor). Glycerine is a viscous liquid. The viscosity increases below 18 °C, therefore glycerine is often mixed with water to improve the handling characteristics. Moreover the added water is not 'free water' that is lost during the production process, so that an end product with a higher moisture content can be sold; that's 'free feed'. Glycerine is sweet, it has a positive effect on feed intake. As a sweetener it can replace molasses in feeds.

Because of economics the usage of glycerine (1,2,3 propanetriol) has been limited as a replacement for propylene glycol (1,2 propanediol) during the transition period of fresh dairy cows. Through the process of gluconeogenesis it's converted in glucose increasing lactose and milk production. Research will have to show if glycerine addition to lactating sows will likewise increase milk production. It can be further hypothesised that preventing glucose deficiencies can decrease fat mobilisation and extensive weight loss during lactation. At the end of the gestation period glycerine can aid the energy supply to unborn piglets if it passed via the placenta. This generates interest for studying the long term effect of glycerine additions to sows on reproduction parameters and vitality of new born piglets.

The energy content of glycerine

The caloric value of glycerine is 397 kcal/mol glycerine, this yield a Gross Energy content of 18.06 MJ/kg glycerine. Figuring a 100% absorption of glycerine the (potential) Metabolisable Energy content is also 18.06 MJ/kg. The Net Energy value will be 14 MJ/kg glycerine, based on an ATP yield of 22 mol ATP/mol glycerine (Bartelt and Schneider, 2002)

The role of glycerine in the energy metabolism

In Figure 2 the possible routes for the metabolism of glycerine are given.

The most important aspects for consideration are:
1. Glycerine needs to be enzymatically activated before it can be metabolised. This initial step is probably limiting and varies among

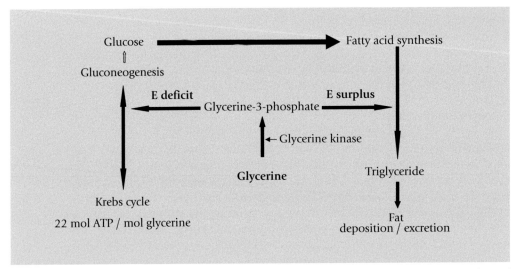

Figure 2. Schematic presentation of metabolism of glycerine (Adapted from Bartelt and Schneider, 2002).

species (see Table 4). As a consequence glycerine accumulates in the plasma after absorption from the gut and is excreted via the urine. This leads to a loss of energy.

2. During (a period of) energy deficiency glycerine can be completely oxidised to CO_2 and H_2O. This yields 22 mol ATP/mol glycerine. Under practical conditions the question remains to what extend glycerine is used as an energy source for animals in a positive energy balance or for growing animal.

3. During (a period of) energy or glucose deficiency glycerine will be converted into glucose via gluconeogenesis. Glucogenic nutrients are needed for lactose production. Glycerine has shown to increase milk yield in dairy cows by increasing propionic acid (a glucogenic nutrient) production in the rumen. By preventing the use of glucogenic amino

Table 4. Dose dependant energy content of glycerine.

Species	Swine (NE MJ/kg)		Layers (AME MJ/kg)		Broilers (AME MJ/kg)	
Glycerine	Pure	80%	Pure	80%	Pure	80%
5%	14.0	11.2	18.1	14.4	18.1	14.4
10%	11.5	9.2	17.3	13.8	17.3	13.8
15%	8.5	6.8	14.9	11.9	15.7	12.6

acids for gluconeogenesis, glycerine additions can increase protein utilisation for e.g. milk protein synthesis.

4. With energy surplus glycerine promotes fat synthesis. This fat can either be deposited as body fat or excreted as milk fat. Glycerine can via gluconeogensis be converted in glucose and further be used for fat production via *de novo* fatty acid synthesis. The energetic efficiency of this process is about 78%. Glycerine can also be used directly as a lipogenic nutrient for triglyceride synthesis by esterifying three fatty acids to it. The energetic efficiency by which absorbed fats are deposited as body fat is 92%. The Net Energy value of glycerine as a glucogenic nutrient is 14 MJ/kg, as a lipogenic nutrient it will than be 16.7 MJ/kg. With an excess of dietary energy mainly derived from starch and sugars or from added fatty acids glycerine is most likely to be used as a lipogenic nutrient, resulting in a very efficient utilisation.

In conclusion the (Net) Energy value of glycerine depends on:
a. the total amount of glycerine given;
b. the animal specie;
c. the physiological status of the animal; and
d. the type of ration (fat, fibre or starch rich).

The use of glycerine in swine and poultry feeds

Kijora *et al.* (1993) found in two experiments a positive effect of the sweet taste of glycerine on feed intake and average daily gain by addition levels up to 10%. With higher addition levels (20 and 30%) the feed intake decreased. The growth declined linearly with inclusion rates higher than 10% (see Figure 3).

Simon *et al.* (1996) found that 10% added glycerine had a positive effect on the average daily gain of broilers. In a follow up study a positive effect of glycerine on feed intake was measured (Simon *et al.*, 1997). In these experiments glycerine additions raised the plasma glycerine concentration significantly from 0.65 to 4.36 µmol/ml and the breast muscle content from 0.4 to 7.5 µmol/g, but liver concentrations were only raised marginally from 16 to 19 µmol/g. It was further mentioned that 26% of the glycerine was excreted via the faeces. A dose response relation was not reported, although up to 25% glycerine was added. Apparently glycerine was metabolised in the liver, preventing accumulation in the liver. However the metabolism was not sufficient to prevent accumulation in tissues and blood resulting in excretion by the kidney.

Bartelt and Schneider (2002) found that with increasing glycerine additions (from 5-15%) the glycerine excretion via urine and excreta increased in a dose depending matter with layers, broilers and pigs (see Figure 4). With a 5% addition the glycerine losses were low. Although

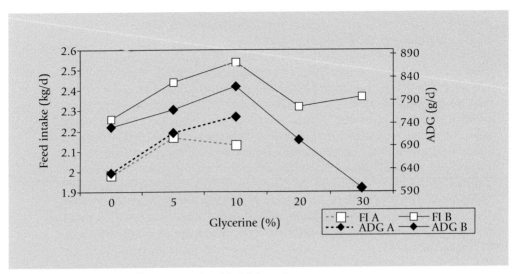

Figure 3. Glycerine feeding to growing/finishing pigs.

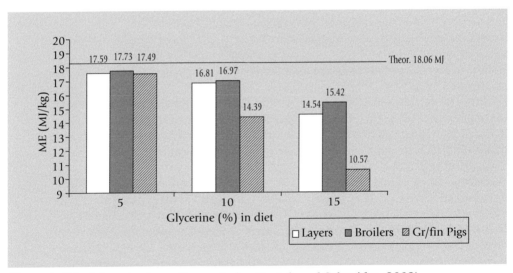

Figure 4. ME of glycerine in poultry and pigs (Bartelt and Schneider, 2002).

at high doses (15%) the ileal digestibility of glycerine was high (> 97%) the excretion was significant. Glycerine is therefore efficiently absorbed from the intestine but at high intake levels the metabolic utilisation is reduced. Possibly the enzymatic activation of glycerine by glycerol kinase to glycerine-3-phosphate is a limiting step. The problem is more profound with pigs than poultry.

In Figure 4 the decrease of the Metabolisable Energy content of glycerine is given with different inclusion rates and animal species. The Metabolisable Energy system takes urinary losses in consideration.

From this figure can be concluded:

a. the utilisation of glycerine is lower with swine than with poultry;
b. the utilisation of glycerine is slightly higher with broilers than layers;
c. the metabolic utilisation of glycerine decreases with higher doses; and
d. for an optimal usage not more than 5% glycerine should be added to swine diets and not more than 10% to poultry diets. The efficiency will than be about 96%.

Apparently there is a threshold level to which glycerine can be metabolised. This threshold level is specie dependant, if it can be influenced directly or indirectly via nutrition (vitamins, minerals) or hormones is unknown.

In broilers and poultry the acetic acid production due to fermentation increased in the small intestine with increasing glycerine additions. At the highest dose (15%) the moisture content of the excreta increased significantly. Increasing excretion levels of glycerine with poultry therefore reduces the litter and (cosmetic) egg quality. Notice that these trials were conducted with pharmaceutical grades of glycerine. Feed grades with a high salt content can aggravate litter problems at lower inclusion rates.

On the basis of these results the energy content of glycerine can be estimated for different species and inclusion rates. For swine a Net Energy value of *pure glycerine* of 14 MJ/kg is used and for poultry an AME of 18.06 MJ/kg.

Glycerine and carcass quality

In two trails in France glycerine addition to swine diets improved the water binding capacity of meat and meat products. Mourot *et al.* (1993) added 5% for the whole feeding period (35-102 kg), while Cerneau *et al.* (1994) added the same amount glycerine only in the finishing period (80-110 kg). In the first trail the drip losses were reduced with 0.5%, in both trials the cooking losses were significantly reduced.

No glycerine effects on average daily gain, lean meat or fat content of the carcass were found. However the fatty acid pattern of inter and intra muscular fat was altered. The amount of saturated (C16:0) and mono unsaturated fatty acids (C18:1) from the *novo* synthesis increased significantly at the expense of the poly unsaturated fatty acids (linoleic (C18:2) and linolenic (C18:3) acid). It can therefore be concluded that even low additions of glycerine for a limited period can significantly improve carcass quality of pork and meat products.

Table 5. Carcass quality of pork.

Glycerine %		0	5
Trial 1 35-102 kg	Drip losses	1.81[a]	1.31[b]
	Cooking losses	30.09[a]	26.57[b]
Trail 2 80-110 kg	Cooking yield	100%[a]	101.9%[b]

[a, b] (P<0.05).

References

Bartelt, J. and Schneider, H.D. 2002 Glycerine in der Tierernährung. UFOP-Schriften, Bonn Germany, 15-36 pp.

Borggreve, G.J. and Veldman, A. 1989. SFR VMD 27 and 28: Rapeseed meal in growing/finishing pig feeds. Schothorst Feed Research report # 268, internal publication.

Cerneau, P., Mourot, J. and Peyronnet, C., 1994. Effets du glycerol alimentaire sur la qualite de la viande de porc et le rendement technologique du jambon cuit. Journées de la Recherche Porcine en France 26: 193-198.

CVB, 2004. Veevoedertabel (Feedstuff table, nutritional value of feed ingredients). Centraal veevoederbureau, Lelystad, the Netherlands.

DeClerq, D.R., 2005. Quality of Western Canadian canola, 17. Canadian Grain Commission, Winnipeg, CA

Fledderus, J., Borggreve, G.J. and Bikker, P., 2000. SFR VML 39. The effect of a high inclusion rate of palmkernel meal, rapeseed meal or sunflower meal in pig feeds. Schothorst Feed Research report # 560, internal publication.

Hickling, D., 2001. Canola meal feed industry guide 2001, 13. Canadian International Grains Institute, Canadian Council of Canada, Winnipeg, CA

INRA, 2004. Table of composition and nutritional value of feed materials. D. Sauvant, J.-M Perez and G. Tran (editors). Wageningen Academic Publishers.

Kijora, C., Bergner, H., Kupsch, R.D. and Hagemann, L., 1995. Glycerol as a feed component in fattening pigs. Arch. Tierernährung 47 (4): 345-360.

Kostic, M.A. and Dart, R.C., 2003. Rethinking the toxic methanol level. Journal of Toxicology: Clinical toxicology 41 (6): 793-800.

Mourot, J., Aumiatre, A., Mounier, A., Peiniau, P., Francois, A., Peyronnet, C. and Jamet, J.P., 1993. Effets du glycerol alimentaire sur les performances de croissance et la qualite de la viande chez le porc large white. Journées de la Recherche Porcine en France 25: 29-36.

Prabhakaran, V, Ettler, H. and Mills, A., 1993. Methanol poisoning: two cases with similar plasma concentrations but different outcomes. Journal of the Canadian Medical Association 148 (6) 981-984.

Rick, M., 2006. The European biodiesel story and it's by products. Biofuels:implications for the feed industry Nutrition conference, Ede, the Netherlands.

Schuman, W, 2005. Glucosinolatgehalt von in Deutschland erzeugten und verarbeiteten Rapssaaten und Rapsfuttermittelen. UFOP-Schriften Heft 27, Bonn Germany.

Simon, A., Bergner, H. and Schwabe, M., 1996. Glycerol as a feed ingredient for broiler chickens. Arch. Tierernährung 49 (2): 103-112.

Simon, A., Schwabe, M. and Bergner, H., 1997. Glycerol supplementation in broiler with low crude protein content. Arch. Tierernährung 50 (3): 271-282.

Südekum, K-H., 2006. Byproducts from biodiesel production: glycerine and rapeseed products in rations for farm animals. In: Biofuels, implications for the feed industry. Wageningen Academic Publishers, The Netherlands.

Südekum, K-H. and Schröder, A., 2002. Glycerine in der Tierernährung. UFOP-Schriften Heft 17, Bonn Germany, pp. 37-67.

Tripathi, M.K. and Mishra, A.S., 2007. Glucosinolates in animal nutrition: a review. Animal Feed Science and Technology 132: 1-27.

Utilization in swine diets of distillers dried grains derived from corn or wheat used in ethanol production

J.F. Patience[1], P. Leterme[1], A.D. Beaulieu[1] and R.T. Zijlstra[2]
[1]Prairie Swine Centre, Box 21057, 2105 8th Street East, S7H 5N9 Saskatoon, Saskatchewan, Canada
[2]University of Alberta, Agricultural, Food and Nutritional Science, T6G 2P5 Edmonton, Canada

Keywords: corn DDGS, wheat DDGS, dry milling process, inclusion in swine diets, carcass quality, nutrient profile

Introduction

The rapid growth in the transportation biofuels sector has attracted considerable attention, not only within the grain sector, but also animal agriculture. A doubling of the portion of the US corn crop directed to ethyl alcohol, or ethanol, production in the past decade, and the expectation that it could double again within the coming decade, has created concern regarding the availability of corn, wheat and other grains for use in livestock diets in the future. Global ethanol production exceeded 44 million litres in 2005, with the US and Brazil each producing about a third of that total (Renewal Fuels Association, 2005a). China, India, France and Germany are other nations producing more than 400 thousand litres of ethanol per year. Germany's production grew at the fastest rate - 60% - between 2004 and 2005; while US growth was only 21%, it experienced greater growth in shear volume in ethanol production than any other nation worldwide and by a very large margin.

Historically, livestock agriculture has competed with the human food market for raw materials; the entry into the marketplace of another large and powerful competitor, growing at a very, very rapid rate, is indeed cause for concern. The livestock sector addressed competition from the human food complex in a number of ways. It produced competitive products (meat, milk and eggs) that were desired by the consumer. It utilized some ingredients that could not enter the human food system due to quality concerns, related in some cases to nutrient profile and in other cases to visual appeal (e.g., colour, kernel size and shape, etc). The livestock sector also utilized co-products of the human food industry. Finally, the quantitative demand for raw grains was such that the livestock industry could co-exist by using that portion of the annual crop that was not required by the human food sector.

The approaches undertaken by the livestock industry to co-exist with the human food sector may only apply in part in the case of this new competition from the transportation biofuels complex. The shear volume of grain required to feed a rapidly growing ethanol industry will reduce the quantity of grain available for livestock. Perhaps just as critical is the pace of growth in the ethanol industry, which demands a significant adjustment by the livestock sector in a very short period of time.

However, like the human food complex, the biofuels industry will provide large quantities of a potentially valuable co-product – distillers dried grain with solubles (DDGS). However, like co-products from the human food sector, DDGS can be highly variable, limiting its utility in pig and poultry diets.

A lack of experience with DDGS, at least in the quantities currently available, combined with early reports of variability in product quality both within and among ethanol plants, has lead to a considerable degree of resistance to their use by some people within the pork and feed industries. In order for wheat- and corn-DDGS to achieve a preferred status as an ingredient in commercial diets, at least three broad technical topics must be addressed:

- well-defined nutrient profiles of the product must be made available, including energy and bio-available amino acids, and to a lesser extent, minerals and vitamins;
- reduced variability, or at least the ability to manage variability in a simple, practical manner;
- palatability for the pig.

Of course, cost and consistent supply will also be important, but these are marketing as opposed to technical issues.

Co-products of ethanol production

In order to maximize the effective utilization of co-products of the ethanol industry, it is useful to understand the manufacturing process. Modern ethanol production is largely achieved by what is called 'dry milling.' Wet milling can also be used to produce ethanol, but most new plants, at least in North America, are based on dry milling.

In dry milling, the entire grain kernel is first ground through hammer mills into flour and processed as such; no attempt is made to separate out specific components of the grain. The ground grain is then mixed with water and enzymes to convert the starch to dextrose. Ammonia is added at this point to control pH and to support yeast growth. The mixture is heated in a high-temperature cooker to kill bacteria, then cooled and transferred to fermenters where the yeast is added to convert the dextrose to ethanol and carbon dioxide. This part of the process requires about two days to complete, after which the material is moved to distillation columns to separate the ethyl alcohol from the remaining 'stillage.' At this point, the alcohol is 190 proof; following further dehydration, it achieves 200 proof concentration. The remaining stillage is centrifuged to separate the solid particles (distillers dried grains or DDG) from the solubles; after evaporation to increase the solids content to about 30%, the solubles are now known as condensed distillers solubles or CDS. The DDG and CDS fractions are commonly reblended to produce distillers dried grains with solubles (DDGS, or sometimes DDG/S) and dried, generally using drum driers (Renewable Fuels Association, 2005b).

According to the Association of American Feed Control Officials, the final DDGS must be named according to the majority grain used in its manufacture. Thus, DDGS produced from 51% corn and 49% other grain or grain would still be called corn DDGS. Because the feedstock used in the production of ethanol can impact the nutrient profile of the resulting DDGS, the specific grains used by a supplier must be confirmed (Gibson and Karges, 2006).

It can be expected that ethanol manufacturers will view co-products differently in the future than in the past. They could be viewed, for example, as an energy source that can reduce the plant's requirement for purchased natural gas. If the CDS are used in this manner, then the by-product from the plant will become simply distillers dried grains, with lower feeding value for pigs. Alternatively, the value of such components as the germ could result in their being removed prior to the fermentation process, and sold as such, also resulting in a lower value DDG by-product for the pig. In short, it is entirely possible that the future will see increased specialization of by-product output from different ethanol plants, as owners seek to improve overall operational efficiency and financial performance.

Nutrient content

Having an accurate nutrient profile for an ingredient is an essential requirement, if that ingredient is to be used effectively in practical diet formulation. One of the current challenges associated with the use of corn- or especially wheat-DDGS is the difficulty in obtaining a complete nutrient profile. This is demonstrated in Table 1 that summarizes information from a number of published feed ingredient databases. Information on wheat DDGS is incomplete.

It is perhaps not surprising that published databases lack completeness with respect to the nutrient composition of DDGS, due to the fact that they were not as important a feed ingredient in swine diets until about 5 years ago. A review of nutrient profiles presented in the literature provides a much more complete picture, at least for corn DDGS (Table 2) because the information is more current. Table 2 provides a summary of recent studies showing the nutrient profile of corn and wheat DDGS.

Table 1. Nutrient profiles of corn and wheat-distillers dried grains with solubles, as reported in published nutrient databases (90% DM)[1].

Source	Corn				Wheat		
	1	2	3	4	2	3	4
Chemical content							
Crude Protein, %	26.8	25.1	-	21.6	33.8	-	23.9
Crude Fat, %	8.1	4.0	-	-	6.5	-	-
Crude Fibre, %	-	7.4	-	7.2	9.2	-	5.4
ADF, %	15.8	9.2	-	-	14.6	-	-
NDF, %	33.5	32.0	-	-	37.9	-	-
Starch, %	-	11.7	-	6.8	3.8	-	3.2
Energy							
DE, Mcal/kg	3.10	2.76	-	3.24	2.70	-	2.92
ME, Mcal/kg	2.73	2.56	-	3.10	2.51	-	2.77
NE Mcal/kg	2.00	1.71	-	1.98	1.60	-	1.69
Amino acids, Total %							
Lysine	0.60	0.63	0.74	0.67	1.05	0.74	0.79
Methionine	0.48	0.44	0.51	-	0.63	0.51	-
T.S.A.A.	0.99	0.93	0.98	0.74	1.44	1.17	0.88
Threonine	0.91	0.88	0.99	-	1.16	1.00	-
Tryptophan	0.24	0.17	0.21	-	0.51	0.33	-
Amino acids, TID %[2]							
Lysine	0.35	0.37	-	-	-	-	-
Methionine	0.36	0.33	-	-	-	-	-
T.S.A.A.	0.67	0.63	-	-	-	-	-
Threonine	0.59	0.55	-	-	-	-	-
Tryptophan	-	0.05	-	-	-	-	-
Macrominerals, %							
Calcium	0.19	0.21	-	-	0.33	-	-
Phosphorus, total	0.75	0.86	-	-	0.67	-	-
Phosphorus, avail.	0.57	0.27	-	-	-	-	-
Sodium	0.24	0.55	-	-	0.05	-	-
Chloride	0.19	0.33	-	-	0.19	-	-

[1] All data presented after adjusting to a common 90% dry matter
[2] Standardized ileal digestible amino acids are presented in data from INRA, 2002.
Source: 1) NRC, 1998; 2) INRA, 2002; 3) Degussa, 2006; 4) Beyer et al., 2003.

Table 2. Nutrient profile of distillers dried grains with solubles derived from corn or wheat (90% DM)[1].

Source	Corn				Wheat	
	1	2	3	4	5	6
No. samples	118	10	9	34	3/2	1
Dry Matter, %	88.9	88.9	90.5	89.2	93.1	91.9
Crude Protein, %	27.2	27.8	26.8	27.8	34.2	40.1
Crude Fat, %	9.8	-	-	9.8	3.3	2.6
Crude Fibre, %	7.9	-	-	6.5	-	6.8
ADF, %	14.6	11.0	14.3	11.0	18.6	19.0
NDF, %	37.9	40.6	34.9	-	32.6	27.3
Starch, %	-	6.6	-	-	-	-
Ash, %	5.2	-	4.8	5.4	4.7	4.8
DE, Mcal/kg	3.59[2]	3.20	-	3.67	3.02	3.62
ME, Mcal/kg	3.37[2]	-	-	3.43	-	-
NE Mcal/kg	-	-	-	1.67	-	-
Amino acids, total, %						
Lysine	0.77	0.80	0.70	0.82	0.62	0.65
Methionine	0.50	0.62	0.51	0.58	-	0.62
T.S.A.A.	-	-	1.03	1.17	-	1.49
Threonine	1.02	1.00	1.02	1.03	1.27	1.15
Tryptophan	0.23	0.17	0.19	0.22	-	0.40
Leucine	3.20	3.15	3.31	3.20	2.48	2.71
Isoleucine	1.01	1.04	0.99	1.05	1.23	1.43
Valine	1.35	1.43	1.34	1.41	1.58	1.72
Histidine	0.68	0.78	0.72	0.76	0.73	0.89
Arginine	1.08	1.12	1.05	1.19	1.44	1.59
Phenylalanine	1.32	1.35	1.44	1.35	1.72	1.94
Macrominerals, %						
Calcium	0.05	-	-	0.06	0.14	-
Phosphorus, total	0.80	0.80	-	0.69	0.85	0.83
Phosphorus, avail.	-	-	-	-	-	0.44
Potassium	0.85	-	-	0.91	-	-
Magnesium	0.30	-	-	0.27	-	-
Sodium	0.22	-	-	0.16	-	-
Chloride	-	-	-	0.08	-	-

[1] All data presented after adjusting to a common 90% dry matter
[2] Calculated from chemical composition; not determined in vivo
Sources: 1) Spiehs *et al.*, 2002; 2) Stein *et al.*, 2006; 3) Cromwell *et al.*, 1993; 4) Shurson, 2006. 5) Nyachoti *et al.*, 2005; 6) Widyaratne and Zijlstra, 2007.

In other words, we anticipate that as common databases are updated, they will contain more complete information on DDGS.

There is a rapidly growing literature on the subject of nutrient composition of corn DDGS; the availability of these data will certainly enhance the use of corn DDGS. However, it is equally clear that insufficient data exists in the published literature to define the nutrient composition of wheat-based DDGS with the same confidence and precision; more research is clearly required. Additional information is likely available from individual ethanol producers, but this information is not as easily obtained.

Energy and carbohydrates

It can be inferred from the data in Tables 1 and 2 that DDGS derived from wheat contains more protein and less fat than DDGS derived from corn. It would similarly appear that fibre levels are similar between the two. It would be helpful to have additional information on the content of remaining starch, if any, in the two products, as this would provide insight into the respective energy values; many energy values in databases and in the literature were derived mathematically, and supportive in vivo data is desirable. The extreme variation in energy concentration of both corn- and wheat-DDGS indicates that additional research is necessary to clarify whether this variation is biological or analytical in nature.

Protein and amino acids

The higher protein content of wheat DDGS presumably reflects the nature of the feedstock, since wheat has more protein than corn. The higher protein content could have implications for the environmental impact of utilizing high levels of DDGS in practical diets (see below), but the major concern to most nutritionists is the content of individual amino acids and their biological availability to the pig. Two common databases (NRC, 1998; INRA, 2002) indicate substantially lower lysine digestibility in corn DDGS than the native corn. The same can be concluded from manuscripts published by Fastinger and Mahan (2006) and Stein et al. (2006). The data of Fastinger and Mahan (2006) is less definitive, since no corn samples were used to compare to the DDGS (Table 3); however, using typical digestibilities for amino acids in corn, such as those published by the NRC (1998), one would draw the same conclusion, albeit with less certainty! The one exception, reported by Stein et al. (2006), was observed with tryptophan, which was actually more digestible in DDGS than in corn. Nyachoti et al. (2005) and Widyaratne and Zijlstra (2007) reported that the amino acids in wheat DDGS are less digestible to the pig than those from wheat grain (Table 3) and that protein in the specific

Table 3. Comparison of the ileal digestibility of amino acids between corn and corn-derived DDGS and wheat and wheat-derived DDGS.

Source	Corn				Wheat			
Reference	1		2		3		4	
Ingredient	Grain	DDGS	Grain	DDGS	Grain	DDGS	Grain	DDGS
No. samples	1	10	-	5	1	2	1	1
Method[1]	Standardized		Standardized		Apparent		Standardized	
Lysine	67.7	56.8	-	51.2	76.1	43.8	78.3	64.1
Methionine	84.0	79.3	-	80.7			88.4	84.2
T.S.A.A.	-	-	-				88.4	81.7
Threonine	73.7	67.0	-	67.7	72.3	62.9	85.0	77.5
Tryptophan	72.8	77.4	-	75.4			91.2	85.7
Isoleucine	77.5	71.7	-	75.9	84.8	68.0	89.1	81.4
Leucine	85.2	80.3	-	82.4	86.7	77.8	88.9	84.5
Valine	76.8	70.6	-	73.7	81.6	67.1	92.4	87.4

Sources: 1) Stein *et al.*, 2006; 2) Fastinger and Mahan, 2006; 3) Nyachoti *et al.*, 2005; 4) Widyaratne and Zijlstra, 2007.

wheat DDGS samples was damaged considerably, based on 25% of the nitrogen existing as non-protein nitrogen.

Other

The phosphorus content of DDGS is also of interest, since phosphorus is an expensive essential nutrient in the pig's diet and since over-formulation can potentially have environmental consequences. Not surprisingly, the phosphorus content of DDGS is higher than in the native grain. Furthermore, the digestibility of phosphorus in DDGS is higher than in raw grains, presumably due to partial destruction of phytate during ethanol production or subsequent processing of the stillage. Widyaratne and Zijlstra (2007) reported that the phytate content of wheat DDGS was considerably lower than in wheat (0.81 versus 1.39%), although the total phosphorus content increased from 0.40 to 1.10% (Table 4).

Table 4. Phytate and phosphorus profile of corn, wheat + corn and wheat distillers dried grains with solubles (% DM).

Variable, %		Distller's Dried Grains with Solubles		
	Wheat	Corn	Wheat + Corn	Wheat
Phosphorus	0.40	0.86	1.02	1.10
Inositol diphosphate (IP2)	0.00	0.00	0.00	0.08
Inositol triphosphate (IP3)	0.00	0.09	0.09	0.09
Inositol quadraphosphate (IP4)	0.00	0.19	0.18	0.28
Inositol pentaphosphate (IP5)	0.00	0.45	0.33	0.64
Phytate (IP6)	1.39	0.92	0.62	0.81

Source: Widyaratne and Zijlstra, 2007.

Nutrient variability

The use of co-product ingredients in swine diets is not a new phenomenon. Experienced nutritionists are well aware that co-products are often more variable in their nutrient composition than the native substrate. Thus, it is no surprise that one of the concerns associated with the use of DDGS in livestock diets is their variable composition and quality.

However, the degree of variability of DDGS may be considerably higher than some other co-product ingredients. The rapid growth of the transportation biofuels industry has resulted in many new operators of ethanol manufacturing plants; the growth and novelty of the industry, as well as the relatively high rates of return previously available from the ethanol alone with minimal revenue generated from the sale of DDGS, have contributed to a poor reputation of this potentially valuable feed ingredient. This situation might change in the future, as the technology associated with ethanol production is improving, resulting in potentially higher quality DDGS (Shurson *et al.*, 2004). Also, margins from ethanol production are declining, so producers may be more inclined to maximize the value of the co-product in livestock feeding. Also, the industry is rapidly gaining experience in the production of high quality DDGS, with some companies already seeking to brand their product as superior to that typically available in the marketplace (Gibson and Karges, undated).

Variability in the nutrient composition of DDGS can be attributed to a multitude of factors, but the most significant are 1) the degree of variation in the incoming corn, 2) proportion of condensed distillers solubles added to the distillers dried grains to produce DDGS, 3) the efficiency with which starch is fermented to ethanol, and 4) the temperature and duration of drying (Shurson, 2004).

Cromwell *et al.* (1993) related an objective measure of colour of DDGS, using the Hunterlab colour procedure (McNaughton *et al.*, 1981), to chick performance. They observed a high correlation between performance and Hunter L (lightness/darkness) and b (yellowness) scores. This suggests that heating, and more specifically overheating, contributes substantially to the differences in performance observed among DDGS samples. In this particular experiment, pig growth rate varied from 390 g/d on the higher quality DDGS to only 218 g/d on the lowest quality samples. These differences in growth rate, observed when corn DDGS represented only 20% of the total diet, are extremely important in modern pork production where barn throughput is a singular priority.

Energy

Variation in energy content is the major concerns associated with the use of significant amounts of DDGS in swine diets. Stein *et al.* (2006) reported that the DE content of 10 samples of corn DDGS varied from 3.81 to 3.38 Mcal/kg DM, a range of 429 kcal DE/kg DM. The data in Table 1 and 2 further demonstrate the variability in the energy content of corn DDGS; an overall range from 2.76 to 3.67 Mcal DE/kg indicates that the data available in the public literature varies widely, due perhaps as much to sampling and methodology of determining energy content as to the ingredient itself. Nonetheless, even within study, the variation in energy content is larger than most nutritionists would prefer, and much greater than in corn. Unfortunately, there is less information available on wheat DDGS, but that available suggests that variation is also a concern.

The likely sources of variation in the energy content of DDGS is 1) the residual starch content, 2) the oil content and 3) the quantity and type of fibre present. The residual starch content is significant, because not only will residual starch contribute to a higher energy content, but it also reflects the degree to which other nutrients in the grain have been concentrated in the DDGS. Less starch suggests that other nutrients, such as fibre, have been more concentrated than might otherwise be the case.

The problem of variability in the energy content of DDGS may also be due, in part, to the manner in which the energy content was determined. In some cases, energy was determined in classical balance studies with pigs, but in other instances, it was not measured directly, but rather was estimated using prediction equations. Depending on the accuracy of the equation(s) when applied to DDGS, this difference in methodology could account for some of the differences in DE or NE observed among studies.

Amino acids

Lysine is obviously the amino acid of greatest interest, because it is first limiting in most practical diets, but in reality, the bioavailable content of all essential amino acids is important to today's nutritionist. Stein *et al.* (2006) reported that the standardized ileal digestibility of lysine averaged 56.8% across 10 samples of corn DDGS, but varied from a low of 43.9% to a high of 63.0%. Fastinger and Mahan (2006) determined an average lysine digestibility of 51.2% across 5 samples, with a range from 38.2 to 61.5%.

These two reports used different methodologies to determine standardized amino acid availability. One adjusted apparent ileal digestibility results using data obtained by feeding a protein-free diet, while the other used a mathematical approach. Keeping this in mind, we can crudely estimate the variability in lysine availability by combining the two studies, resulting in 15 observations. On this basis, the mean lysine availability was 55.0% and the standard deviation was 7.6%. Since obtaining large numbers of results from ileal collection studies is expensive, this 'estimate' may be useful to nutritionists seeking to adopt a statistical approach to dealing with ingredient variability. Comparable data (mean; standard deviation) for the other amino acids would be methionine (79.8%; 3.6%), threonine (67.2%; 4.1%) and tryptophan (76.7%; 3.5%). Recently, Stein (2006) reported a mean standardized ileal digestibility of lysine in 37 corn DDGS samples of 62.3% with a standard deviation of 7.6%; this mean is higher but the standard deviation is identical to the one calculated above. Corresponding means and standard deviations for methionine, threonine and tryptophan were 81.9%/4.1%, 70.7%/5.3% and 69.9%/7.0%, respectively.

Of course, the lysine data could over-estimate true lysine bioavailability if the drying process has resulted in the production of Maillard products. The ileal collection method fails to fully identify reductions in actual biological availability due to heat damage.

There is much less data available in the public literature on amino acid availability in wheat DDGS, but the study reported by Nyachoti *et al.* (2005) indicates that the problem is similar to that observed in corn DDGS.

Other

The problem of variability does not appear to be as large with phosphorus as it is with energy and amino acids. The level of phosphorus in both wheat and corn DDGS appeared to be relatively consistent across databases (Table 1) and individual studies (Table 2).

Feeding trials

A number of experiments have been undertaken to estimate the level of DDGS that can be included in practical diets without impairing pig performance. Whitney and Shurson (2004) fed up to 25% corn DDGS to weanling pigs; based on these results, about 10% corn DDGS could be included in diets for pigs between 8 and 25 kg bodyweight. Some impairment in growth might occur in the initial period on test, but results for the overall experiment showed no effect on performance when 10% DDGS was included in the diet.

Whitney *et al.* (2006) included up to 30% corn DDGS in diets fed to pigs from about 28 to 115 kg, and observed that 10% was a fairly safe level in terms of growth performance; however, loin depth, belly thickness and belly firmness were all reduced at this level. At 20% inclusion, a modest decline in growth rate was observed and carcass quality further declined, with lower dressing percent, smaller loins, and elevated iodine number in belly fat. Thacker (2006) recently reported similar findings with wheat DDGS; compared to the control diet, as little as 5% DDGS in the grower diet and 3% in the finisher diet resulted in carcasses with smaller loins and lower dressing percentage. In this particular study, impairment in carcass quality was the same at low levels of DDGS inclusion as compared to much higher levels (25%), raising questions as to the cause of the problem.

The DDGS market appears to be differentiating, with some suppliers developing and marketing what they consider a superior product with greater consistency in nutrient composition than 'generic' product. Gowans *et al.* (2007) reported feeding up to 25% of a 'high quality" corn DDGS to pigs from 35 to 90 kg with no adverse affect on overall performance.

Feed intake often is a challenge, at least with wheat DDGS. Widyaratne and Zijlstra (2007) reported that 25% wheat DDGS in the diet of pigs from 50 to 85 kg reduced feed intake with no impact on feed efficiency; thus, growth rate declined in proportion to the drop in feed intake. The reason for the decline remains unclear. It could be palatability, inaccurate formulation for energy or amino acids, concentration of an anti-nutritional factor in the DDGS, or some residue of the ethanol manufacturing process left in the DDGS. Further research is underway to study this problem.

Ultimately, economics will play a significant role in determining the inclusion rate of DDGS in swine diets. However, because barn throughput is an important issue on North American pig farms, research is critical to overcoming the decline in growth rate observed at higher levels of DDGS inclusion, if this product is to achieve significant penetration in the swine feed market.

Other issues

Carcass fat quality

Distillers grains derived from corn will contain quite high levels of fat, in the range of 8 to 10% or more. The fat is highly unsaturated, so that feeding corn DDGS raises questions regarding the degree of unsaturation of carcass fat. Typically, an iodine value for carcass fat of less than 70 is preferred (Barton-Gade, 1987). The impact of dietary fat on the iodine value of the carcass can be estimated from the following equation (Pettigrew and Esnaola, 2001):

$$IV_{CARCASS} = 47.1 + 0.14 \text{ X } [(IV_{DIETARY FAT} \text{ X Fat Content of the Diet})/10]$$

Different markets have different standards for belly fat firmness, but suffice to say that the quantity of corn DDGS used in pig diets could be limited by a need to maintain a maximum iodine value in the carcass. There will be much less concern with wheat DDGS, which has a much lower fat content – about 1/3 that of corn.

Phosphorus

Another nutrient concentrated in the DDGS is phosphorus. In corn or wheat grain, the total phosphorus content is 0.25% and 0.34%, respectively (Patience et al., 1995) and of this, about 30% is digestible (INRA, 2002). Thus, these grains contribute between 0.08% and 0.10% available phosphorus. However, like other nutrients, phosphorus has been concentrated in the DDGS, such that corn and wheat DDGS contain about 0.80% total phosphorus (Tables 1 and 2). Widyaratne and Zijlstra (2007) determined that measured the phosphorus in corn and wheat DDGS is between 50% and 55% digestible, meaning that they contribute 0.40% or more of digestible phosphorus, about 4 times that of the native grain. Since the digestible phosphorus content of DDGS is substantially higher than that required by pigs weighing more than 10 kg (NRC, 1998), the quantity of DDGS in the diet could be limited by the need to avoid excessive phosphorus in the slurry. This will depend on the other ingredients present in the diet and their available phosphorus content.

Current recommendations

At the present time, based on variability in nutritional quality and growth performance data, commodity DDGS produced from corn can safely be used at 10 to 15% of the diet of growing and finishing pigs;

corn DDGS of certified quality might be included at higher levels. The use of corn DDGS in weanling diets is less certain, but 7% to 10% is likely acceptable, based on admittedly limited data.

The acceptable level of wheat DDGS is 5 to 10% in diets of growing and finishing pigs. Higher levels may be possible, but only when the cause of the feed intake depression can be ascertained.

In terms of carcass and meat quality, the quantity of DDGS that can be used will be market and genotype dependent. Clearly, the issue is the iodine value of carcass fat; thus, the acceptable level of corn DDGS will be very much market-dependent.

References

Barton-Gade, P.A., 1987. Meat and fat quality in boars, castrates and gilts. Livestock Prod. Sci. 16:187-196.

Beyer, M., Chudy, A., Hoffmann, L., Jentch, W., Laube, W., Nehring, K. and Schiemann, R., 2003. Rostock Feed Evaluation System. Reference numbers of feed value and requirement on the base of net energy. Plexus Verlag, Miltenberg-Frankfurt. 392 pp.

Cromwell, G.L., Herkelman, K.L. and Stahly, T.S., 1993. Physical, chemical and nutritional characteristics of distillers dried grains with solubles for chicks and pigs. J. Anim. Sci. 71:679-686.

Degussa, 2006. AminoDat 3.0 Degussa AG, Hanau.

Fastinger, N.D. and Mahan, D.C., 2006. Determination of the ileal amino acid and energy digestibilities of corn distillers dried grains with solubles using grower-finisher pigs. J. Anim. Sci. 84:1722-1728.

Gibson, M.L. and Karges, K., 2006. DDG/S production: Present and future. Proc. Midwest Swine Nutrition Conference. Indianapolis, IN. pp. 42-57.

Gibson, M.L. and Karges, K., 2006. Use of distillers grains for swine. www.dakotagoldmarketing.com/ UserFiles/File/Swine%20Research/DDGS%20or%20Swine%20NC%20Nut%20Conf%20-%20102605.pdf. Accessed December 27, 2006.

Gowans, J., Callaahan, M., Yusupov, A., Campbell, N. and Young, M., 2007. Determination of the impact of feeding increasing levels of corn dried distillers grains on the performance of growing-finishing pigs reared under commercial conditions. Proc. Banff Pork Seminar, Banff AB. pp. A22.

INRA, 2002. Tables of Composition and Nutritional Value of Feed Materials. Wageningen Academic Publishers, Wageningen.

NRC, 1998. Nutrient Requirements of Swine. 10th ed. National Academy Press, Washington, DC.

McNaughton, J.L., Reese, F.N and Deaton, J.W. 1981. Relationship between color, trypsin inhibitor contents, and urease index of soybean meal and effects on broiler performance. Poult. Sci. 60:393.

Nyachoti, C.M., House, J.D. Slominski, B.D. and Seddon, I.R., 2005. Energy and nutrient digestibilities in wheat dried distillers' grains with solubles fed to growing pigs. J. Sci. Food Agric. 85:2581-2586.

Patience, J.F., Thacker, P.A. and de Lange, C.F.M., 1995. Swine Nutrition Guide, Prairie Swine Centre, Saskatoon. 274 pp.

Pettigrew, J.E. and Esnaola, M.A., 2001. Swine nutrition and pork quality: A review. J. Anim. Sci. 79: E316-E342.

Renewable Fuels Association, 2005a. Industry Statistics: www.ethanolrfa.org/industry/ statistics/#E. Accessed December 27, 2006.

Renewable Fuels Association, 2005b. How ethanol is made: www.ethanolrfa.org/resource/ made. Accessed December 27, 2006.

Shurson, G., 2006. Nutrient profiles: Comparison tables: www.ddgs.umn.edu/profiles/ us-profile%20comparison %20tables%20January%202006.pdf. Accessed December 27, 2006.

Shurson, G., Spiehs, M. and Whitney, M., 2004. The use of maize distiller's dried grains with solubles in pig diets. Pig News and Information. 25(2):75N-83N.

Spiehs, M.J., Whitney, M.H. and Shurson, G.C., 2002. Nutrient database for distiller's dried grain with solubles produced from new plants in Minnesota and South Dakota. J. Anim. Sci. 80:2639-2645.

Stein H.H., 2006. DDGS: Energy and nutrient content and digestibility. Proc. Midwest Swine Nutr. Conf., Indianapolis, IN. pp. 58–66.

Stein, H.H., Gibson, M.L. Pedersen, C. and Boersma, M.G., 2006. Amino acid and energy digestibilities in ten samples of distillers dried grain with solubles fed to growing pigs. J. Anim. Sci. 84:853-860.

Thacker, P.A., 2006. Nutrient digestibility, performance and carcass traits of growing-finishing pigs fed diets containing dried wheat distillers grains with solubles. Can. J. Anim. Sci. 86:527-529.

Whitney, M.H. and Shurson, G.C., 2004. Growth performance of nursery pigs fed diets containing increasing levels of corn distiller's dried grains with solubles originating from a modern Midwestern ethanol plant. J. Anim. Sci. 82:122-128.

Whitney, M.H., Shurson, G.C. Johnson, L.J. Wulf, D.M. and Shanks, B.C., 2006. Growth performance and carcass characteristics of grower-finisher pigs fed high-quality corn distillers dried grains with solubles originating from a modern Midwestern ethanol plant. J. Anim. Sci. 84:3356-3363.

Widyaratne, G.P. and Zijlstra, R.T., 2007. Nutritional value of wheat and corn distiller's dried grain with solubles: digestibility and digestible contents of energy, amino acids and phosphorus, nutrient excretion and growth performance of grower-finisher pigs. Can. J. Anim. Sci. (in press).

Distillers dried grains plus solubles for ruminants

L.L. Berger
University of Illinois

Abstract

Distillers dried grains plus solubles (DDGS) is a co-product resulting from the conversion of corn to ethanol. Distillers grains are an excellent source of nutrients for ruminants. Distillers grains are a good source of protein (> 30% CP) that is high in ruminally undegradeable protein (ROP), and a good source of energy (NE_L = 2.25 Mcal/kg of DM) for lactating cows. Most research suggests it has 110-130% the energy of corn for growing and finishing cattle. As a by-product, variability in composition is a potential concern when feeding DDGS. Nutrients concentrations vary with year, agronomic conditions, corn variety, plant equipment and processing conditions, varying ratios of distillers grains and distillers solubles being combined, drying conditions, etc. The University of Wisconsin researchers recently published an excellent review of the use of DDGS for lactating cows. They summarised 21 trials with 53 comparisons of DDGS and control diets fed to 690 cows per treatment. They divided the inclusion rates of DDGS into low (5-15%), medium (16-25%) and high (26-48%) levels, respectively. As a result of the meta-analyses these researchers concluded that DDGS is an excellent source of nutrition for lactating cows. Dry matter intake responses are most favourable when DDGS is fed at 5-15% of the diet. Milk yield responses are affected by dietary crude protein, forage source and amino acid balance. Growing and finishing cattle offer the greatest opportunity for the utilisation of DDGS in the United States. We conducted a trial to determine the optimum level of wet and dry distillers grains for dairy-beef steers based on performance and carcass value. The control diet was 70% high-moisture corn, 15% corn silage, and 15% of a soybean meal-based supplement. The soybean meal and corn were replaced with 12.5, 25.0, 37.5 and 50% dry distillers, or 25, 37.5 and 50% wet distillers grains (DM basis). Steers fed the control diet tended to have the lowest dry matter intakes, but the best feed efficiencies. Rate of gain average about 1.4 kg/day for all treatments except for steers fed 50% wet distillers which tended to be slightly lower. Dry matter intakes were greater ($P < 0.01$) for steers fed the dry distillers compared to the wet distillers. In contrast, feed efficiency was superior ($P < 0.01$) for steers fed the wet vs. the dry distillers grains.

Keywords: ruminants, distillers grains, nutrients, lactation, growth

Introduction

Agriculture in the United States is going through dramatic changes that could not have been envisioned just a few years ago. Traditionally agriculture has been the source of food and fiber, but now it is seen as a source of liquid fuel, ethanol. Growth in the number of ethanol plants has been so rapid that some experts are predicting that by 2015, ethanol production could reach 15 billion gallons/year (56.78 billion liters/yr). If this becomes reality, it will result in the production of approximately 40 million metric tons of distillers grains plus solubles (DDGS) per year. Plants that currently exist or are under construction will produce approximately 22 million metric tons of distillers grains by the end of 2007. Although many factors such as a drop in crude oil price, reduce ethanol price, drought, etc, could slow the conversion of corn to ethanol, all experts agree that the supply of DDGS will increase dramatically.

Nearly all the growth in ethanol production is via the dry milling process resulting in DDGS. The wet milling of corn resulting in corn gluten feed, is expected to stay fairly constant. Some dry milling plants are removing the germ to extract the oil for bio-diesel resulting the production of a modified distillers grains. Additional changes in process technology are likely to increase the number and variability of co-products available in the future.

Nutrient profle

Distillers grains are an excellent source of nutrients for ruminants. Distillers grains are a good source of protein (> 30% CP) that is high in ruminally undegradeable protein, and a good source of energy (NE_L = 2.25 Mcal/kg of DM) for lactating cows. Most research suggests it has 110-130% the energy of corn for growing and finishing cattle. The 12% fat and readily digestible fiber (35-40% NDF) contributes to it high energy content. It is a very palatable ingredient that is readily accepted by all classes of ruminants. Common inclusion rates in the U.S. vary from 10 to 50% of the diet dry matter.

As a by-product, variability in composition is a potential concern when feeding DDGS. Nutrients concentrations vary with year, agronomic conditions, corn variety, plant equipment and processing conditions, varying ratios of distillers grains and distillers solubles being combined, drying conditions, etc. The Dairy One feed analysis laboratory in Ithaca, New York provides the best public access database to measure nutrient variation. This laboratory receives feed samples from all over the United States. The data has not been screened to remove outliers, so the standard deviations are probably higher than what would be found in other data sources. The Dairy One dried DDGS data base is robust in that it contains 1,500-2,000 samples for most nutrients that were analysed between May of 2000 and June 1 of 2006. All nutrients are expressed on a dry matter basis and can be accessed at http://www. dairyone.com/Forage/FeedComp/disclaimer.asp.

In this data set there were 2,234 samples that averaged 30.29% crude protein with a standard deviation of 3.64 (Table 1). This is similar to the 1996 Beef NRC data that averaged 30.4% crude protein and a standard deviation of 3.55. According to these data sets the normal range (means ± one standard deviation) of CP for DDGS is 26.7 to 33.9%. The wet DDGS data set gives similar values.

Acid detergent insoluble crude protein (ADICP) is used to estimate the unavailable protein. Samples that are over-dried are expected to contain more ADICP. The 1,882 samples analysed averaged 4.92 percentage

Table 1. Distillers grains plus solubles nutrient composition: current vs. 1996 beef national research council values.

Item	Dairy One[a]			1996 Beef NRC[b]		
	N	Mean	Std. Dev[c].	N	Mean	Std. Dev.
Dry matter, %	2310	87.28	9.90	450	90.3	2.00
Crude protein, %	2234	30.29	3.64	439	30.4	3.55
Acid detergent insoluble crude protein, %	1882	4.92	2.07			
Neutral detergent insoluble crude protein, %	535	9.70	3.84			
Acid detergent fiber, %	1870	16.79	3.91	370	21.3	4.82
Neutral detergent fiber, %	1845	33.49	4.87	158	46.0	8.71
Ether extract, %	1601	12.95	2.91	166	10.7	3.12
Ash, %	601	5.92	1.18	18	4.6	0.86
Calcium, %	1501	0.092	0.14	384	0.26	0.23
Phosphorus, %	1518	0.918	0.14	384	0.83	0.15
Magnesium, %	1495	0.319	0.05	383	0.33	0.08
Potassium, %	1495	1.069	0.19	383	1.08	0.27
Sodium, %	1248	0.196	0.19	382	0.30	0.26
Sulfur,%	1047	0.648	0.18	113	0.44	0.15
Zinc, ppm	1244	64.58	37.70	383	67.8	23.9
Copper, ppm	1244	5.05	22.73	383	10.6	7.81

[a]http://www.dairyone.com.
[b]1996 Beef NRC.
[c]Standard deviation.

units of crude protein as ADICP with a standard deviation of 2.07. It interesting to note that in the same feed analysis library, 967 samples of wet distillers grains plus solubles averaged 3.43 percentage units of crude protein as ADICP with a standard deviation of 2.28. A difference of 1.5 percentage units of crude protein as ADICP and similar standard deviations between the wet and DDGS, suggest that only a small reduction in protein availability results from drying. There were no ADICP given in the 1996 NRC.

The NDF (33.49 vs. 46.0%) and ADF (16.79 vs. 21.3%) are considerably lower for this data set than what was reported in the 1996 Beef NRC (Table 1). Part of the explanation may be due to a much larger number of samples for NDF (1,845 vs. 158) and ADF (1,870 vs. 370) with lower standard deviations in the Dairy One data set.

With 1,602 samples, the crude fat averaged 12.95% with a standard deviation of 2.91. This is considerably higher than the 10.7% (N=166) reported in the 1996 Beef NRC. The cause of this increased fat concentration is not known. Variable fat levels can be especially important when formulating diets for lactating cows. High levels of unsaturated fat can cause a depression in fiber digestion and milk fat percentage.

Phosphorus concentrations on 1,518 DDGS samples averaged 0.918% with a standard deviation of 0.14% (Table 1). This is also higher than 0.83% average for the 383 samples in the Beef NRC. Phosphorus concentrations may determine inclusion rate in many diets where nutrient management of the waste is a problem. Considerable research is underway to develop methods of phosphorus removal from DDGS. However at present, there are no low-phosphorus DDGS products on the market.

Sulfur concentrations can also be important when feeding high levels of DDGS to ruminants. The 1,047 samples averaged 0.648% sulfur with a standard deviation of 0.18. This would make the normal range of sulfur concentrations between 0.47 and 0.83%. High sulfur concentrations in ruminant diets can increase the risk of polioencephalomalacia. In addition, high sulfur levels can bind several minerals, especially copper. Copper concentration average 5.05 ppm for the 1,244 samples analysed. The standard deviation is suspicious at 22.7, when it is compared with the wet DDGS that averaged 3.75 ppm with a standard deviation of 5.7. Special attention should be given to the copper concentrations in diets containing high levels of DDGS.

Mycotoxins

Mycotoxins concentrations are of interest because they can be highly variable and are concentrated approximately three-fold during the fermentation process. Generally, mycotoxins are not a problem in corn in the upper-Midwestern part of the United States where most of the ethanol plants are located. It is difficult to find large data sets where mycotoxins concentrations in DDGS have been measured. In the Dairy One data set it appears that concentrations are similar between wet and dry distillers grains. Aflatoxin concentrations averaged 3.6 ppb when measured in a total of 60 wet and dry distillers samples with standard deviations of 2.4 and 5.0 (Table 2). A total of 65 wet and dry distillers samples average 1.9 ppm of vomitoxin. Only 11 samples were analysed for zearaleonone and averaged 0.55 ppm. Fusmonisin concentrations were determined in 40 samples and averaged 0.60 ppm. Based on this relatively small data set and practical feeding experience, it appears that mycotoxin concentrations are not a major concern with DDGS for ruminants.

Table 2. Mycotoxin concentrations in wet and dry distillers grains plus solubles based on dairy one analyses.

Item	Wet distillers[a]			Dry distillers[b]		
	N	Mean	Std. Dev[c].	N	Mean	Std. Dev.
Aflatoxin, ppb	28	4.83	2.39	22	2.39	5.04
Vomitoxin, ppm	40	2.28	1.73	25	1.49	1.40
Zearalenone, ppm	6	0.31	0.23	5	0.78	0.08
Fumonisin, ppm	19	0.69	1.22	21	0.51	0.11

[a]http://www.dairyone.com.
[b]1996 Beef NRC.
[c]Standard deviation.

Within plant variation

These data suggest there is considerable variation in nutrient profile across the corn fermentation industry. However, the questions many users are asking is how much reduction in variation can I achieve if I get DDGS from one plant. In an attempt to answer that question we worked with a 48-million gallon (181 million liters) per year plant in the upper-Midwest. For over two years they randomly sampled a load leaving their plant on a monthly basis. A private independent laboratory did the nutrient analysis and the results were sent directly to a University of Illinois representative.

Most of the nutrients concentrations are similar to what was reported by Dairy One, but the potential variation is still striking. The crude protein averaged 31.05%, with a standard deviation of 2.05 and a minimum and maximum of 27.17 and 35.31%, respectively (Table 3). The ADICP averaged 4.14%, with a standard deviation of 1.86 and a range of 1.78 to 11.38% for the minimum and maximum, respectively. It appears that at least one load was over-dried causing approximately a third of the protein to be bound to the ADF. The NDF and ADF concentrations averaged 34.31 and 17.06%, with standard deviations of 4.21 and 3.11, respectively. The crude fat concentrations tended to be high for this plant. Crude fat averaged 15.24%, with a standard deviation of 1.55 and minimum and maximum values of 12.22 and 19.01%, respectively. This variation in fat level could create challenges if the DDGS was fed at high concentrations in the diet. Phosphorus concentrations averaged 0.89% with a standard deviation 0.11 and minimum and maximum values of

Table 3. Within plant variation of distillers grains plus solubles monitored monthly over two years.

Item	Mean	Std. Dev.[a]	Minimum	Maximum
Dry matter, %	88.27	2.18	84.16	92.77
Crude protein, %	31.05	2.05	27.17	35.31
Acid detergent insoluble CP, %	4.14	1.86	1.78	11.38
Neutral detergent insoluble CP, %	6.95	2.72	3.55	14.15
Neutral detergent fiber, %	34.31	4.21	26.59	46.36
Acid detergent fiber, %	17.06	3.11	7.98	26.53
Ether extract, %	15.24	1.55	12.22	19.01
Starch, %	6.46	2.60	1.80	12.31
Calcium, %	0.06	0.16	0.02	1.17
Phosphorus, %	0.89	0.11	0.28	1.02
Magnesium, %	0.38	0.03	0.27	0.45
Potassium, %	1.23	0.21	0.81	2.39
Sulfur, %	0.55	0.17	0.11	0.95
Sodium, %	0.31	0.16	0.03	0.64
Chloride, %	0.19	0.17	0.11	0.97
Sugar, %	3.73	2.73	0.01	12.7

[a] Standard deviation.

0.28 and 1.02%, respectively. It is difficult to understand how a sample of DDGS could contain only 0.28% phosphorus because corn averages 0.30% phosphorus. Sulfur concentrations were also quite variable. Mean sulfur concentrations were 0.55%, with a standard deviation of 0.17 and minimum and maximum values of 0.11 and 0.95%, respectively.

Many of the samples in this data set were taken shortly after start-up when the management was focused on the fermentation steps and less concerned about quality control of the DDGS. However, the ethanol production is always going to be first priority because at current ethanol prices about $6 of revenue are generated from ethanol for every dollar generated from the sale of DDGS.

Utilisation by ruminants

Ruminants can make efficient use of DDGS because it is a good source of digestible fiber, rumen undegradeable protein and energy. Ruminal *in situ* NDF disappearance for wet and dry distillers grains was 4.4 and 3.7%/hr (Firkins *et al.*, 1985). This was similar to the 5.1 and 4.7%/hr disappearance of NDF from wet and dry corn gluten feed, respectively. In contrast rates of crude protein degradation were markedly different for DDGS and corn gluten feed. Rates of in situ crude protein disappearance were, 3.90, 4.27, 9.46 and 8.93%/hr for wet and dry distillers grains and wet and dry corn gluten feed, respectively. Rates of crude protein disappearance for DDGS was about half of the soybean meal control at 9.97%/hr. When these corn proteins were fed to duodenually cannulated steers the ruminal undegraded protein was 47, 54, 26, and 14% for wet and dry distillers grains and wet and dry corn gluten feed, respectively (Firkins *et al.*, 1984).

Dairy cattle

University of Wisconsin researchers recently published an excellent review of the use of DDGS for lactating cows (Kaiser *et al.*, 2005). They summarised 21 trials with 53 comparisons of DDGS and control diets fed to 690 cows per treatment. They divided the inclusion rates of DDGS into low (5-15%), medium (16-25%) and high (26-48%) levels, respectively. The forage base varied among trials resulting in 30 comparisons with > 50% corn silage, 16 comparisons with > 50% haycrop, 7 comparisons of a 50:50 mix. Crude protein concentrations were divided into low (<16%), medium (16-18%), and high (>18%).

Compared to the control diets, low medium and high levels of DDGS resulted in a –0.17, 0.60, and –0.97 kg/day response in dry matter intakes. The intake response was greater with corn silage (-0.86) than with haycrop (0.20) or mixed forage (0.13) diets. The 3.5% fat-corrected milk yield response to DDGS was 0.65, 0.19, and −1.43 kg/day for the low, medium and high inclusion rates, respectively. Crude protein concentrations affected the response to DDGS inclusions. The 3.5% fat-corrected milk yield response to feeding DDGS was −1.18, -0.19 and 0.87 kg/day for cows fed low, medium and high levels of crude protein, respectively.

As a result of the meta-analyses these researchers concluded that DDGS is an excellent source of nutrition for lactating cows. Dry matter intake responses are most favorable when DDGS is fed at 5-15% of the diet. Milk yield responses are affected by dietary crude protein, forage source and amino acid balance. Many dairymen are feeding 2.0 to 3.0 kg DDGS per head daily to the Holstein cows.

Growing cattle

Growing and finishing cattle offer the greatest opportunity for the utilisation of DDGS. For example, dairy-beef (Holstein steers) often enter the feedlot at 130 kg and leave 300 days later at 600 kg and can consume nearly a 1,000 kg of DDGS. With this in mind, we conducted a trial to determine the optimum level of wet and dry distillers grains for dairy-beef steers based on performance and carcass value. The control diet was 70% high-moisture corn, 15% corn silage, and 15% of a soybean meal-based supplement. The soybean meal and corn were replaced with 12.5, 25.0, 37.5 and 50% dry distillers, or 25, 37.5 and 50% wet distillers grains (DM basis). Steers were fed their respective diets for 270 days, after which they were harvested at a commercial processing facility.

Steers fed the control diet tended to have the lowest dry matter intakes, but the best feed efficiencies (Rincker and Berger, 2003). Rate of gain average about 1.4 kg/day for all treatments except for steers fed 50% wet distillers which tended to be slightly lower. Dry matter intakes were greater ($P<0.01$) for steers fed the dry distillers compared to the wet distillers. In contrast, feed efficiency was superior ($P<0.01$) for steers fed the wet vs. the dry distillers grains. Carcass dressing percentage was higher ($P<0.05$) for steers fed the distillers grains than the control diet. There was no difference in carcass quality grade, yield grade or hot weight due to dietary treatment except for steers fed the 50% wet distillers had the lightest carcasses and the least back fat.

When the dietary treatments were compared on a profit per head basis, it became obvious that the corn to DDGS price ratio will determine the optimum level of feeding. In this data set, when DDGS price was greater than 110% of the corn price, 12.5 to 25% inclusion levels were the most profitable. When the price of DDGS equaled corn price, then the 25-30% inclusion levels were best. When DDGS price was less than 90% of the corn price, higher inclusion levels became the most profitable.

In summary, DDGS is an excellent source of nutrition for ruminants. It is high in digestible fiber, undegradable protein, and energy. The variation in nutrient concentration offers some challenges that need to be considered. Its high phosphorus and sulfur concentrations may determine inclusion levels in some production environments. The optimal inclusion level will be determined by the price of competitive protein and energy sources. In the U.S., DDGS are being utilised successfully at 5-50% of the diet for ruminants.

References

Dairy One, http://www.dairyone.com/Forage/FeedComp/disclaimer.asp (accessed 10-27-2006).

Firkins, J.L., Berger, L.L. and Fahey Jr., G.C., 1985. Evaluation of wet and dry distillers grains and wet and dry corn gluten feeds for ruminants. J. Anim. Sci. 60:847-860.

Firkins, J.L., Berger, L.L., Fahey Jr., G.C., and Merchen N.R., 1984. Ruminal nitrogen degradability and escape of wet and dry distillers grains and wet and dry corn gluten feeds. J. Dairy Sci. 67:1936-1944.

Kaiser, R., Schuab, E., Shaver, R. and Armentano L., 2005. Using Ethanol Industry Co-products in Dairy Rations. http://www.wisc.edu/dysci/uwex/nutritn/presentn/presentn.htm. (accessed 12-27-2006).

NRC, 1996. Nutrient Requirement of Beef Cattle. 7th ed. National Academy Press, Washington, DC.

Rincker, C.B. and Berger L.L., 2003. Optimizing the use of distiller grain for dairy-beef production. Distiller Grain Trial, University of Illinois.

Keyword index

Printed in the United States
by Baker & Taylor Publisher Services